ACID MINE DRAINAGE

ACID MINE DRAINAGE

Sustainable Mining Practices for Mitigating Environmental Impact

ATHANAS SIMON MACHEYEKI

Department of Geology, College of Earth Sciences and Engineering, The University of Dodoma, Dodoma, Tanzania

NOEL ISACK KAAYA

Government Chemist Laboratory Authority (GCLA), Dodoma, Tanzania

MATHIAS ATHANAS MACHEYEKI

Earth Sciences Institute of Shinyanga (ESIS), Shinyanga, Tanzania

ELSEVIER

Elsevier
Radarweg 29, PO Box 211, 1000 AE Amsterdam, Netherlands
50 Hampshire Street, 5th Floor, Cambridge, MA 02139, United States

ISBN: 978-0-443-45107-2

For Information on all Elsevier publications visit our website at https://www.elsevier.com/books-and-journals

Publisher: Peter Llewellyn
Acquisitions Editor: Jennette McClain
Editorial Project Manager: Prachi Mittal
Production Project Manager: Sathyavani Deiva Sigamani
Cover Designer: Bakyalakshmi S

Typeset by Aptara, New Delhi, India

Contents

Biography

Athanas Simon Macheyeki is a Professor of Geology at the University of Dodoma. He previously served as Commissioner of the Minerals Commission of Tanzania, Executive Secretary of the Tanzania Extractive Industries Transparency Initiative, Manager of Applied Geology at the Geological Survey of Tanzania, and Principal of the Mineral Resources Institute in Dodoma. He is the Founder of the Earth Sciences Institute of Shinyanga. Prior to his service in the Government of Tanzania, he worked as a Mineral Exploration Geologist with Anglo American Tanzania. He holds a BSc (University of Dar Es Salaam), MSc (University of the Western Cape), and PhD (Ghent University). He developed lithogeochemical ratios for the exploration of Ni–Cu sulfide deposits, including indices such as Pd/V × 1000 (Macheyeki, 2011). He has authored and coauthored more than 30 scholarly publications, including journal articles, geological maps, book chapters, and books, notably *Applied Geochemistry* (Macheyeki et al., 2020) and *The East African Rift System* (Macheyeki & Kafumu, 2024).

Noel Isack Kaaya is a Chemist at the Government Chemist Laboratory Authority, Tanzania, where he specializes in environmental monitoring, chemical safety, and regulatory compliance. His work involves the analysis of environmental samples from mining and industrial areas to assess compliance with environmental and public health standards. He also oversees the production, transportation, use, and market surveillance of chemicals to minimize risks to human health and the environment. His extensive engagement with artisanal and small-scale gold mining (ASGM) operations has fostered a strong interest in environmental geochemistry and contaminant assessment. His MSc research focused on the geochemical characteristics of ASGM tailings and their associated ecological and human health risks. His research has resulted in peer-reviewed publications on trace-element contamination, ecological and human health risk assessment, and the recovery of rare earth elements from mining wastes. In addition to his practical expertise in environmental and elemental analysis, his research interests encompass environmental geochemistry, contaminant fate and transport, ecological risk assessment, human biomonitoring, environmental epidemiology, and hydrological modeling of environmental contaminants. He holds a BSc in

Chemistry from the University of Dar es Salaam and an MSc in Chemistry from the University of Dodoma, Tanzania.

Mathias Athanas Macheyeki is a cybersecurity and digital forensics engineer and a full-stack software developer. He has experience in building secure, practical, and user-centered digital systems, conducting vulnerability assessments, penetration testing, secure code reviews, and incident response activities. He also has experience in machine learning and artificial intelligence technologies and has developed several solutions within this field. He is the co-founder of TEYORA Technologies (teyora.co.tz) and Nexor Digital Limited, technology companies focused on cybersecurity, digital forensics, artificial intelligence, and software development. His work involves web and mobile application development, network security, and digital forensics, with a focus on reliability and real-world usability. Mathias has collaborated with the Earth Science Institute of Shinyanga to design and implement a secure Student Records Management System and has supported the Geological Survey of Tanzania through system maintenance, software installations, and technical troubleshooting.

References

Macheyeki, A. S. (2011). Lithogeochemistry in exploration for Ni-Cu sulphide deposits in the Kabanga Belt. *Journal of African Earth Sciences*, *61*, 62–81.

Macheyeki, A. S., Li, X., Kafumu, D. P., & Yuan, F. (2020). Application of non-conventional mineral exploration techniques: Case studies. Elsevier: USA, 210p.

Macheyeki, A. S., & Kafumu, D. P. (2024). The East African Rift System: Natural Resource Potentials. Elsevier: USA, 364p.

Foreword

Acid mine drainage (AMD) remains one of the most persistent and challenging environmental legacies of mining worldwide. Characterized by the formation of highly acidic, metal-laden waters, AMD poses significant risks to aquatic ecosystems, human health, and infrastructure. In extreme cases, these leachates can reach pH values as low as −3.6, though they more commonly remain above zero. As a global environmental issue, AMD's long-term impacts are profound—second only to a broader challenge—"the climate change" in scale and persistence. Effective AMD management requires robust regulatory frameworks, but these must be built upon a strong technical foundation.

This book provides a comprehensive and structured introduction to AMD from a technical perspective. It begins with fundamental principles and historical case studies, then systematically presents a range of treatment technologies—from conventional to advanced, active to passive systems. A key feature is the inclusion of worked examples, guiding readers through the calculation and interpretation of critical AMD parameters. This practical approach bridges the gap between theory and real-world application, enhancing both understanding and utility.

What sets this volume apart is its forward-looking, interdisciplinary scope. Dedicated chapters on artificial intelligence (AI) and policy, governance, and community engagement expand the discussion beyond traditional technical boundaries. The AI section explores modern monitoring and modeling approaches and introduces a purpose-built software tool for AMD analysis and prediction. Readers are encouraged to apply this tool with their own data, fostering engagement and iterative improvement. The policy chapter situates AMD within a broader global context, emphasizing governance, stewardship, and community involvement.

By integrating scientific, technological, and socioregulatory perspectives, this book offers a holistic treatment of AMD. It is designed to serve a wide audience—students, researchers, industry professionals, policymakers, and affected communities—providing the knowledge and tools needed to understand, manage, and mitigate AMD impacts.

In an era where sustainable resource development is increasingly critical, this work stands as a timely contribution toward bridging science, practice, and policy in addressing one of mining's most enduring environmental challenges.

Jabulani Ray Gumbo
Director, Research and Innovation,
Bindura University of Science Education,
Bindura, Zimbabwe

Preface

Growing environmental and public health concerns in the Lake Zone (Tanzania) have increasingly been linked to mining-related activities (e.g., Kaaya et al., 2025). Among the potentially toxic elements identified, arsenic constitutes the largest proportion (approximately 86%), followed by nickel and chromium. Recent studies by Kaaya et al. (2026) indicate that chromium and nickel are primarily of geogenic origin, derived from the weathering of mafic rocks, whereas arsenic is largely associated with sulfide-rich gold ores. These ores, when exposed to oxygen and water, can generate acidic conditions (pH values as low as 3.01) and elevated concentrations of dissolved metals, leading to the formation of acid mine drainage (AMD).

Although concerns regarding the occurrence and impacts of AMD in mining regions of Tanzania have been raised over time, comprehensive and accessible scientific resources on this subject remain limited. This book has been developed in response to this gap, with the aim of providing a clear, structured and evidence-based account of AMD that is relevant both locally and globally. The text begins with the fundamental principles of AMD, including its formation, controlling geochemical reactions and pathways of transport in natural systems. It then examines the environmental and human health implications associated with AMD, drawing on examples from diverse geological and mining contexts. A central component of this book is devoted to management, mitigation and remediation strategies. Both conventional approaches (including active treatment systems) and emerging, nonconventional techniques are discussed, with an emphasis on effectiveness, sustainability and applicability across different settings. The book further addresses prediction and monitoring, covering established methods, such as static and kinetic testing, as well as advancing approaches involving real-time monitoring and digital technologies, including artificial intelligence. Practical examples are included to support understanding and application. In addition, this book considers the broader policy, legal and governance dimensions of AMD. It highlights the importance of regulatory frameworks, responsible environmental management, and meaningful community engagement in addressing the challenges associated with mining and its legacy. This book is intended for a wide and diverse readership, including students, researchers, academics, industry practitioners, policymakers, legal professionals and communities affected by mining activities. Every effort has been made to present the material in a manner that is scientifically rigorous while remaining accessible to readers from different backgrounds and disciplines.

As authors, we are deeply aware that the challenges associated with AMD extend beyond science alone—they are environmental, societal and intergenerational in nature. Our motivation for writing this book has been shaped not only by scientific curiosity but also by a sense of responsibility to contribute to sustainable solutions in regions where mining continues to play a vital economic role. We hope that this work will not only inform but also inspire collaboration across disciplines and sectors, encourage responsible stewardship of natural resources and support efforts to safeguard both ecosystems and human well-being. Ultimately, we offer this book as a contribution toward a more informed, inclusive and sustainable future for all communities impacted by mining activities.

Athanas Simon Macheyeki
Noel Isack Kaaya
Mathias Athanas Macheyeki

References

Kaaya, N. I., Vegi, M. R., & Macheyeki, A. S. (2025). Health risks of geogenic contaminants in gold mining areas in Geita, Tanzania. *Journal of Trace Elements and Minerals, 12*(3), 100222. https://doi.org/10.1016/j.jtemin.2025.100222.

Kaaya, N. I., Vegi, M. R., & Macheyeki, A. S. (2026). Acid mine drainage, soil pollution, and carcinogenic risk of geogenic contaminants in artisanal and small-scale gold mining areas of Geita, Tanzania. *Environmental Geochemistry and Health, 48*(1), 23. https://doi.org/10.1007/s10653-025-02909-8.

Acknowledgments

The authors are indebted to Mr. Godson Mulazi and Mr. Baraka Joseph Toha of Geology Department, The University of Dodoma (UDOM) for assisting in data compilation for Chapters 3 and 4. Mr. Sabato Mtesi, a former geology student at UDOM; Mr. Edwin S. Igenge, Head of the Legal Services Unit of the Tanzania Atomic Energy Commission; Mr. Chone Malembo of the Mining Commission of Tanzania together with Mr. Julius Sarota, a retired economic geologist from the then Ministry of Energy and Minerals made significant contributions to Chapter 6. Dr. Petro Karungamye from the College of Natural Sciences and Mathematics at UDOM reviewed the early version of Chapter 4. Eng. Selemani Msangi, a freelance Mining Engineer, commented on Chapter 5. The authors thank Prof. Godlisten Kombe of UDOM for moral support. The materialization of this project would not have been possible, however, without the moral support of our beloved families. We are equally indebted to our respective coworkers.

CHAPTER ONE

Introduction to acid mine drainage

1.1 Introduction

Communities living near metallic and coal mines often face health risks due to acid mine drainage (AMD) and acid rock drainage (ARD)—forms of pollution resulting from mine wastes and sulfide-containing natural rocks, respectively (Dold, 2017; Masindi et al., 2018; Olds et al., 2016). Of these, AMD is particularly harmful, as it triggers chemical transformation from mining areas into soil and water systems, leading to extensive environmental degradation (Worlanyo & Jiangfeng, 2021). The generation of AMD from mining activities poses challenges to humanity due to its intense pollution and difficulty in management. Acid mine drainage can lower pH in the environment, sometimes to negative pH values in extreme conditions. For instance, in Spain, the average pH of −1.56 in the AMD has been reported in open air (Sarmiento et al., 2018). In the United States, severe impacts of AMD on waterways have been reported by the US Geological Survey, with some streams experiencing pH levels as low as −3.6 (Nordstrom et al., 2000). Such conditions result from pyrite oxidation, concentration by evaporation and minor contributions from aqueous ferrous iron oxidation and efflorescent mineral formation (e.g., melanterite), causing ecological damage (Wolkersdorfer et al., 2020). Widespread soil contamination by AMD from mining regions in China has led to severe degradation of agricultural lands and reduction of their productivity (Liu et al., 2018). In Tanzania and other countries south of the equator, AMD poses severe health risks to local communities, although comprehensive data are often unavailable (e.g., Almås & Manoko, 2012). Its toxic effects devastate ecosystems, threaten biodiversity and jeopardize health and livelihoods of communities worldwide (Ighalo et al., 2022; Jiao et al., 2023).

Ighalo et al. (2022) and Jiao et al. (2023) group the environmental impacts of AMD into four types: (1) chemical (such as increased acidity and metals in the environment), (2) physical (such as reduced light penetration in the surrounding water column), (3) biological (such as behavior, respiration,

Acid Mine Drainage: Sustainable Mining Practices for Mitigating Environmental Impact
DOI: https://doi.org/10.1016/B978-0-443-45107-2.00007-1

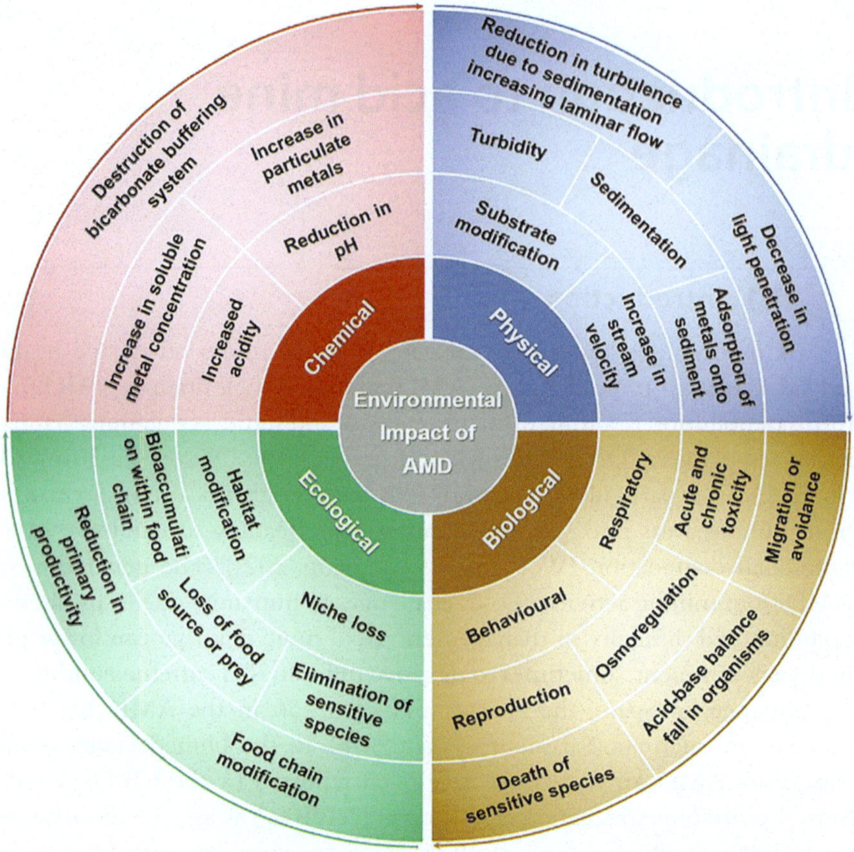

Figure 1.1 The environmental impact of acid mine drainage categorized into four main groups: physical impacts (e.g., reduced light penetration and sedimentation), biological impacts (e.g., loss of sensitive species and migration), chemical impacts (e.g., increased acidity and heavy metal contamination), and economic impacts (e.g., altered food chains and habitats, loss of food sources/prey). *(From Jiao et al. (2023)).*

reproductive defects and mortality of organisms) and (4) ecological (such as altered food chains and habitats as well as loss of food sources/prey) (Fig. 1.1).

Socially, AMD can undermine food and water security, compromise human health, destabilize communities and hinder economic growth (Masindi et al., 2018). However, while treating AMD, circular economic operations, including the recovery of strategic minerals and sulfuric acid, can turn out to be a positive impact on AMD (Akinwekomi et al., 2020; Masindi et al., 2022). Thus AMD is one of the most pressing environmental challenges of our time worldwide (e.g. (Lindsay et al., 2015)). As mining operations expand, the need for sustainable solutions grows ever more urgent. This book takes an in-depth look at the origins, impacts and innovative strategies for combating AMD. Grounded in scientific research and environmental justice principles, it seeks to inspire actionable change—bridging the gap between technical solutions and the social responsibility of safeguarding our planet for future generations. The book focuses on AMD as the primary source of pollution associated with metallic and coal mining activities, exploring its causes, consequences and potential mitigation strategies.

Acid mine drainage does not just harm the environment; it erodes the fabric of communities that depend on clean water and healthy ecosystems. "Acid Mine Drainage: Sustainable Mining Practices for Mitigating Environmental Impact" is the definitive resource for understanding and addressing this critical issue. For professionals, they would gain insights into cutting-edge mitigation techniques and policy frameworks. For advocates, they would learn how to champion environmental justice in impacted regions. For students and academics, they are expected to explore a comprehensive guide that blends rigorous research with actionable solutions.

1.1.1 Understanding AMD

Acid mine drainage is an acidic water produced when sulfur-bearing minerals such as pyrite (FeS_2), pyrrhotite ($Fe_{(1-x)}S$), are exposed to oxygen and water during mining activities (Kusi et al., 2024; Sarmiento et al., 2024). When pyrite, pyrrhotite or other sulfides are exposed to oxygen and water, they form acidic leachate—a fluid often associated with high concentrations of sulfates and dissolved metals such as As, Sb, Cu, Pb, Cd, Zn, Hg, Ag, Sn, Fe, Al, Mn, Tl, U, Th and W (Bhattacharya et al., 2006; Jung et al., 2023; Modabberi et al., 2013). According to Sarmiento et al (2024), one of the effects of acid leachate is on the mechanical properties of carbon steel and aluminum (specifically AISI 1020 carbon steel and AW6060 aluminum) . Sarmiento et al. (2024) further demonstrate how the dynamic process of acid leachate (erosion corrosion due to natural flow of the stream) can highly affect materials faster than in a static environment (chemical corrosion only). They show that for over 80 days of exposure in a static scenario, weight loss is almost 2% for aluminum and 0.5% for steel, fatigue strength is reduced by 28% for steel and ultimate tensile strength is reduced by almost 5% for aluminum and by 2.5% for steel. However, in a dynamic scenario, the weight loss after 4 days of exposure is 30%–35% for both materials, the fatigue strength is reduced by almost 50% for steel and the ultimate tensile strength is reduced by 33% for aluminum and by almost 5% for steel.

Considering the current world's rush on critical minerals, AMD is of particulalr concern (Deberdt et al., 2025; Liu et al., 2023; Saadaoui et al., 2025). The world is in a rush for the so-called critical minerals, which some authors define as (1) minerals needed to facilitate the clean energy transition to carbon net zero by 2050, (2) minerals used in defense and security applications and (3) minerals employed in communication and entertainment technologies (Saadaoui et al., 2025). These include , rare earth elements, graphite, Li, Co, Cu, U and Ni to mention a few (Deberdt et al., 2025; Kaaya et al., 2025). The current rush for critical minerals is therefore driving expansion of mining activities that intensify acid generation processes, thereby increasing long-term environmental and health risks if AMD is not effectively predicted and managed.

1.2 Formation of AMD

Acidification is a global environmental concern arising from the complex interaction of natural processes, human activities and climate change dynamics (Duan et al., 2025; Hassoun et al., 2025). Rainwater, groundwater, surface water bodies, oceans, atmosphere and soils, which are essential for sustaining both aquatic and terrestrial life, are increasingly affected by acidification (Bai et al., 2025; Chen et al., 2026; Dominique et al., 2025; Duan et al., 2025; Eswar et al., 2025; Figuerola et al., 2025; Hassoun et al., 2025; Hsiao et al., 2025; Hulver et al., 2025; Jing et al., 2025; Kang et al., 2025; Liu et al., 2025; Naveed et al., 2025; Olson et al., 2025; Patra & Punekar, 2025; Stewart et al., 2025; Tang et al., 2025; Wang et al., 2025; Zhang et al., 2025, Kaaya et al. 2026; Zhanget et al. 2025).

Understanding the mechanism and processes behind the formation of acidification is essential for developing effective strategies to prevent, minimize, treat and manage its severe, long-term environmental, health and economic impacts (Skousen et al., 2019). Although natural biogechemical processes, such as emission of S and N oxides from volcanic eruptions, natural weathering of sulfide rocks and decomposition of organic matter, produce acidity, human activities are the dominant drivers of global acidification, making their control essential (Rice & Herman, 2012). Rice and Herman (2012) present various mechanisms by which human activities can increase the generation of acids in the environment (Table 1.1).

Table 1.1 Acidification processes exacerbated by human activities and their environmental effects.

Element oxidized	Acid generated	Means of human exacerbation	Environmental effect	Scale of effect	Nature of effect	Potential for natural amelioration	Trends in human exacerbation
S oxidation	Strong acid, H_2SO_4	Fossil-fuel combustion	Acidic atmospheric deposition	Regional	Acidified receiving waters (freshwater); acidified soils	Rate of bedrock weathering too slow to balance stress; depleted exchange capacity in soils	Western hemisphere emission controls implemented and emissions reduced; global fuel consumption increasing
S oxidation	Strong acid, H_2SO_4	Mining coal and base metal sulfides; processing and smelting metal sulfide ores	Acid mine drainage	Local	Acidified receiving waters (freshwater)	Bedrock weathering restricted to geochemically reactive substrate	Global materials and energy consumption increasing

(continued on next page)

Table 1.1 Acidification processes exacerbated by human activities and their environmental effects—cont'd

Element oxidized	Acid generated	Means of human exacerbation	Environmental effect	Scale of effect	Nature of effect	Potential for natural amelioration	Trends in human exacerbation
Fe oxidation and hydrolysis	Free H^+ is released via Fe^{3+} hydrolysis	Mining coal and base metal sulfides; processing and smelting metal sulfide ores	Acid mine drainage	Local	Acidified receiving waters (freshwater)	Bedrock weathering restricted to geochemically reactive substrate	Global materials and energy consumption increasing
C oxidation	weak acid, H_2CO_3	Fossil-fuel combustion; deforestation; cement manufacturing; biofuel development	Elevated CO_2 in atmosphere	Global	Acidification of oceans	Neutralization capacity of oceans inadequate	Global fuel consumption increasing; deforestation increasing
N oxidation generates	strong acid, HNO_3	Fossil-fuel combustion; production and use of N fertilizers	Acidic atmospheric deposition	Regional	Acidified receiving waters (freshwater); acidified soils	Rate of bedrock weathering too slow to balance stress; depleted exchange capacity in soils	Global food production and consumption increasing; transportation sector increasing

Modified after Rice and Herman (2012).

Acid mine drainage forms when sulfide minerals deep in the earth are exposed during mining; highway construction and other large-scale excavations get in contact with water and oxygen (Orndorff & Daniels, 2004; Skousen et al., 2019). In such situations, most sulfide minerals oxidize to form sulfuric acid, metal ions and sulfate, which, if untreated, enter surface- or groundwater (Skousen et al., 2019). The acid produced from oxidation of sulfide minerals is generally neutralized by naturally occurring neutralizers (Modabberi, Alizadegan, & Mirnejad, 2013; Paktunc, 1999; Skousen et al., 1997). A balance between the acid-producing potential and acid-neutralizing capacity of the disturbed overburden will indicate the ultimate acidity or alkalinity that can be expected from water draining from disturbed materials upon complete weathering (Skousen et al., 1997). Many types of acid-neutralizing minerals are present in rocks, including metal hydroxides, carbonate minerals and some silicate minerals, but only carbonates occur in sufficient quantity to effectively neutralize acid-producing rocks (Jiao et al., 2023; Skousen et al., 2019).

1.2.1 Oxidation of pyrite to generate sulfuric acid

Not all forms of sulfur species in the waste rock or tailings produce acid. The iron sulfide minerals (pyrite, pyrrhotite and arsenopyrite) are know to generate acid. Other sulfides yield less acid than pyrite or may not yield acid at all (Hakkou et al., 2008; Mansur et al., 2021). Certain sulphides and sulphates as well as organic sulphur, do not generate acids. Dold (2017) lists the amount of protons (to represent acidity) produced when some sulfide minerals are oxidized (Table 1.2).

Pyrite oxidation can occur by mechanisms, such as (1) chemical processes involving oxygen and / or ferric iron, (2) biological processes by microorganisms and (3) electrochemical processes by galvanic oxidation (Evangelou & Zhang, 1995). Biological processes of pyrite oxidations are mainly the action of microbial organisms on oxidation of sulfur species to sulfuric acid by (1) direct (enzymatic) metabolic reactions and (2) indirect (chemical, Fe^{3+}-mediated) metabolic reactions (Third et al., 2000). The direct metabolic reactions involve the physical attachment of bacterial cell to the mineral surface, followed by enzymatic reactions that break down the pyrite, where pyrite acts as the electron donor for microbial energy metabolism (Fowler et al., 2001). Examples of enzymes for oxidation of sulfur species released by microbes are sulfur oxygenase reductase and quinone oxidoreductase (Bulaev et al., 2025). In the indirect metabolic oxidation

Table 1.2 Overview of the oxidation reactions of the most relevant sulfide minerals via atmospheric oxygen, considering the hydrolysis of ferric iron and the final amount of protons liberated.

Sulfide mineral	Formula	Moles H^+
	Oxidation via oxygen + hydrolysis of Fe^{3+}	
Pyrite (FeS_2)	$FeS_2 + 3.75O_2 + 3.5H_2O = Fe(OH)_3 + 2SO_4^{2-} + 4H^+$	4
Arsenopyrite (FeAsS)	$FeAsS + 2O_2 + 3H_2O = Fe(OH)_3 + SO_4^{2-} + HAsO_4^{2-} + 3H^+$	3
Chalcopyrite ($CuFeS_2$)	$CuFeS_2 + 4O_2 + 3H_2O = Cu^{2+} + Fe(OH)_3 + 2SO_4^{2-} + 2H^+$	2
Pyrrhotite ($Fe_{(1-x)}S$)	$x = 0.1$: $Fe_{(0.9)}S + 2.175O_2 + 2.35H_2O = 0.9Fe(OH)_3 + SO_4^{2-} + 2H^+$	2
Enargite (Cu_3AsS_4)	$Cu_3As_4 + 8.75O_2 + 2.5H\text{-}_2O = 3Cu^{2+} + HAsSO4^{2+} + 4SO4^{2+} + 4H^+$	4
Sphalerite (ZnS)	$ZnS + 2O_2 + = Zn^{2+} + SO_4^{2-}$	0
Galena (PbS)	$PbS + 2O_2 + = Zn^{2+} + SO_4^{2-}$	0
Covellite (CuS)	$CuS + 2O_2 + = Zn^{2+} + SO_4^{2-}$	0
	Oxidation via ferric iron	
Pyrite (FeS_2)	$FeS_2 + 14Fe^{3+} + 8H_2O = 15Fe^{2+} + 2SO_4^{2-} + 16H^+$	16
Arsenopyrite (FeAsS)	$FeAsS + 13Fe^{3+} + 8H_2O = 14Fe^{2+} + SO_4^{2-} + HAsO_4^{2-} + 15H^+$	15
Chalcopyrite ($CuFeS_2$)	$CuFeS_2 + 16Fe^{3+} + 8H_2O = Cu^{2+} + 17Fe^{2+} + 2SO_4^{2-} + 16H^+$	16
Pyrrhotite ($Fe_{(1-x)}S$)	$x = 0.1$: $Fe_{(0.9)}S + 7.8\ Fe^{3+} + 4H_2O = 8.7Fe^{2+} + SO_4^{2-} + 8H^+$	8
Enargite (Cu_3AsS_4)	$Cu_3AsS_4 + 35Fe^{3+} + 20H_2O = 3Cu^{2+} + HAsO_4^{2-} + 35Fe^{2+} + 4SO_4^{2} + 39H^+$	39
Sphalerite (ZnS)	$ZnS + 8Fe^{3+} + 4H_2O = 8Fe^{2+} + SO_4^{2-} + 8H^+$	8
Galena (PbS)	$PbS + 8Fe^{3+} + 4H_2O = 8Fe^{2+} + SO_4^{2-} + 8H^+$	8
Covellite (CuS)	$CuS + 8Fe^{3+} + 4H_2O = 8Fe^{2+} + SO_4^{2-} + 8H^+$	8

After Dold (2017).

of pyrite by bacteria (bioleaching), microorganisms facilitate the chemical oxidation of pyrite without direct physical contact with mineral surface by continuously regenerating the primary chemical oxidant, the ferric ion (Fe^{3+}) from the ferrous ion (Fe^{2+}) produced by chemical reaction (Dash et al., 2025). Under highly acidic conditions, acidophilic bacteria can catalyze the oxidation of Fe^{2+} to Fe^{3+}, and then, oxidation of pyrite by Fe^{3+} becomes the main mechanism whereby acid is produced (Kang et al., 2015). *Thiobacillus ferrooxidans* and *Leptospirillum ferrooxidans* are capable of catalyzing the oxidation of Fe^{2+} to Fe^{3+}. The microbial communities essential for AMD formation are diverse, including prokaryotic (bacteria), eukaryotic microbes and archaea (Qiu et al., 2025). Specifically, the bacteria of AMD can be divided into iron-oxidizing bacteria, sulfur-oxidizing bacteria, sulfate-reducing bacteria (SRB) and acidophilic bacteria (Picard et al., 2018; Duverger et al., 2020).

Despite the negative impacts of AMD brought by microbial pyrite oxidation, the understanding of mechanistic interactions between microbes and minerals has turned out to several key advantages: (1) biomining and bioleaching, enabling recovery of metals such as Cu, Au, Ni and Co from low-grade ores and mine tailings; (2) reduced energy and chemical consumption, as microbial processes replace high-temperature smelting and aggressive chemical leaching; (3) environmental remediation, through the use of SRB in AMD treatment systems to neutralize acidity and precipitate metals; and (4) sustainable resource utilization, by improving the economic viability of low-grade and complex ores and extending mine life.

Chemical oxidation of pyrite involves O_2 and Fe^{3+} oxidants. Oxygen adsorbs on the surface of protonated pyrite, leading to the formation of different products (e.g., oxyhydroxides and hydroxysulfates) dictated by pH (Bonnissel-Gissinger et al., 1998). At circumneutral pH, O_2 is the main oxidant and goethite (α-FeO(OH)) is the dominant precipitate formed (Geysels et al., 2025). In acidic condition, Fe^{3+} is the main oxidant. Ferric ion accepts electrons from pyrite (FeS_2) forming Fe^{2+} (Holmes & Crundwell, 2000). According to Dold (2017), ferric oxidation of pyrite produces more acidity than oxygen oxidation. However, it should be noted that ferric oxidation requires the presence of O_2 to ensure a constant Fe^{3+} supply (Elghali et al., 2023).

The galvanic (or electrochemical) oxidation of pyrite is ensured by the physical contact between an anodic and a cathodic sulfide in a medium that facilitates electron transfer between two minerals. Being a semiconductor

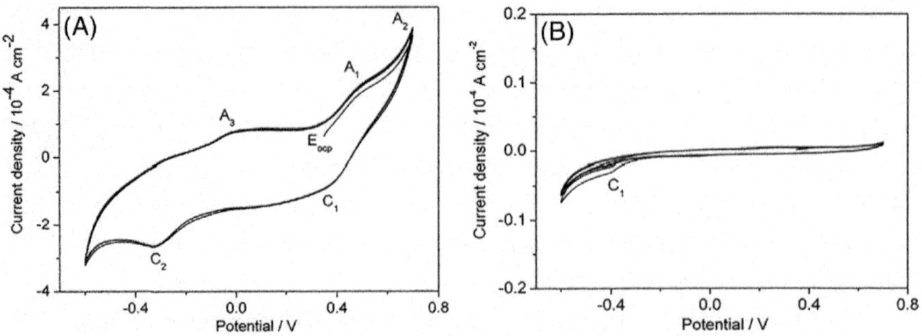

Figure 1.2 ***A cyclic voltammetry (CV) curve for (A) a freshly polished pyrite electrode and (B) S_8 in electrolyte.*** (A) Three anodic peaks labeled A1, A2, and A3; and two cathodic peaks labeled C1 and C2. (B) The four-cycle CV curve of shows extremely low current density and no anodic oxidation peaks, signifying that once formed, S_8 is difficult to oxidize further electrochemically. *(From Tu et al. (2017)).*

(Savage et al., 2008; Liu et al., 2025), pyrite acts as an anode, transferring electrons to an oxidant (O_2 or Fe^{3+}), leading to the formation of iron oxides / hydroxides and sulfate ions, often involving sulfur intermediates such as elemental sulfur (S^0) or polysulfides (S_n^{2-}, where "*n*" is the number of sulfur atoms) (Rimstidt & Vaughan, 2003). This process can be surface-passivated by sulfur layers (S^0 −rich layer at the surface of pyrite) at lower potentials but reactivated at higher potentials, changing the surface chemistry to produce hydrophobic polysulfides, affecting mineral flotation. Tu et al. (2017) employ electrochemical and surface analytical techniques to systematically investigate oxidation of pyrite at pH 2. Their study reveal that electrochemical oxidation of pyrite occur through two reaction pathways where the first pathway occur at low potential (0.50 V) and is characterized by formation of a surface layer of $Fe(OH)_3$ and S^0, which result in surface passivation on pyrite surface and, hence, reduced oxidation. Passivation decrease when the potential is increased to 0.60 V due to conversion of amorphous elemental sulfur (S_8) to crystalline S_8, resulting in the previously covered active sites being reexposed which allow continued oxidation of pyrite (Tu et al., 2017). The second reaction pathway of pyrite oxidation occur at potentials of 0.70 and 0.80 V, resulting in the formation of $Fe(OH)_3$ and SO_4^{2-}, which results in more S_8 and polysulfides ($S\text{-}_n^{2-}$), together with an iron-rich layer composed of $Fe(OH)_3$, FeO and Fe_2O_3, forming and accumulating on the pyrite surface, leading to a decreased rate of oxidation rather than a complete passivation of the surface (Fig. 1.2).

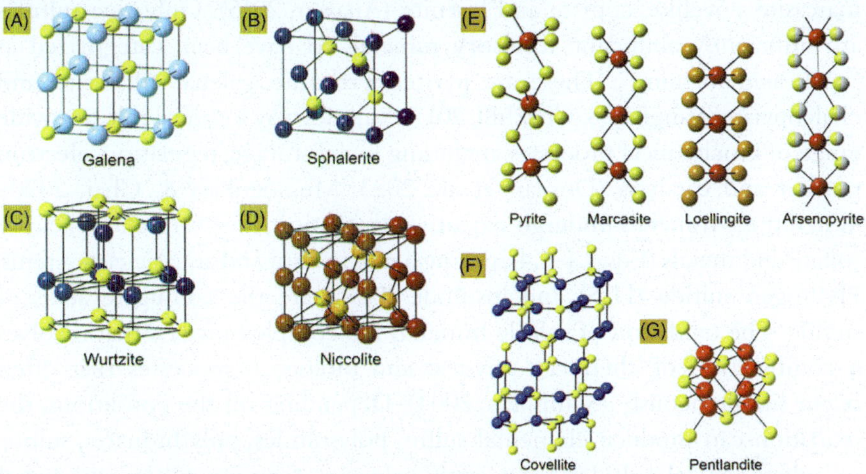

Figure 1.3 Crystal structures of the major sulfides. (A) Galena (PbS). (B) Sphalerite (ZnS). (C) Wurtzite (ZnS). (D) Niccolite (NiAs). (E) Linkage of metal–sulfur octahedra along the c-axis direction in pyrite (FeS2), marcasite (FeS2), loellingite (FeAs2), and arsenopyrite (FeAsS). (F) Covellite (CuS). (G) Cube cluster of tetrahedrally coordinated metals in the pentlandite structure. Atom colours as follows: light blue—Pb; purple—Zn; yellow—S; red—Fe; orange—As; dark red—Ni; dark blue—Cu; for 1E arsenopyrite, yellow/grey spheres are either S or As. *(From Vaughan and Corkhill (2017)).*

1.3 Sulfide minerals

Sulfides are strong nucleophiles that can bind to transition metal ions, forming various metal-sulfide centers with different compositions and structures (Fig. 1.3). A sulfide can also be defined as an inorganic anion of sulfur with the chemical formula S^{2-} or a compound containing one or more S^{2-} ions. According to Vaughan and Corkhill (2017), sulfides are compounds in which sulfur is combined as an anion with a metal (or semimetal) cation or cations. Arsenic or Sb, sometimes Se and Te minerals, can also be included in this definition. The sulfosalts are a special group of the sulfide minerals that have a general formula AmTnXp and in which the common elements are A = Ag, Cu, Pb; T = As, Sb, Bi; X = S. Solutions of sulfosalts are corrosive. These minerals are commonly found in ore and associated with mines of gold, base metals, Hg, U and coal. When sulfide minerals in a rock are disturbed by mining, they react with O_2 and H_2O to produce sulfuric acid. This acid then dissolves heavy metals like Fe, Cu and Pb, which can be toxic to aquatic life and humans. Important sulfides are arsenopyrite, pyrite, pyrrhotite, chalcopyrite, chalcocite, galena, sphalerite, marcasite, covellite, cuprite and bornite (Anawar, 2015). Only five sulfides are sufficiently abundant accessory minerals to have been categorized as "rock acid forming". These are pyrite, pyrrhotite, galena, sphalerite, and chalcopyrite (Vaughan & Corkhill, 2017). Sulfides play a crucial role in a wide range of biochemical processes, including metal storage, regulation, electron transfer and catalysis (Doman et al., 2023; Musaogullari & Chai, 2020). When these minerals undergo oxidative dissolution, they can release acidity, sulfate and metals (Eqs. 1.1–1.4), posing a risk to land and aquatic ecosystems. Hydrogen sulfide (H_2S) and bisulfide (HS^-) are the conjugate acids of sulfide. The oxidation of sulfide minerals is a complex process that involves a combination of chemical, physical and biological processes that often occur simultaneously (Schippers, 2004). Depending on the conditions, the oxidation can produce elemental sulfur, polysulfides, polythionates, sulfite, or sulfate. Metal sulfides react with halogens, forming sulfur and metal salts.

$$8MgS + 8I_2 \rightarrow S8 + 8MgI_2 \tag{1.1}$$

The sulfide ion does not exist in aqueous alkaline solutions of Na_2S (Queizán, Graña, & Hermida-Ramón, 2022). Instead, sulfide converts to hydrosulfide:

$$S^{2-} + H_2O \rightarrow SH^- + OH^- \tag{1.2}$$

Upon treatment with an acid sulfosalts convert to hydrogen sulfide:

$$S^{2-} + H^+ \rightarrow SH^- \tag{1.3}$$

$$SH^- + H^+ \rightarrow H_2S \tag{1.4}$$

Sulfides may be formed from sulfates. For example, approximately half of the sedimentary organic matter in marine coastal environments is metabolized with sulfate as the terminal electron acceptor (Jørgensen, 1982). The electron acceptor sulfate is reduced to sulfide in the process and some of this sulfide becomes sequestered as pyrite. Four sulfides (pyrite, FeS_2; chalcopyrite, $CuFeS_2$, pyrrhotite, Fe_7S_8; and arsenopyrite, FeAsS) shall be discussed here owing to their abundance in natural environments (Schoonen, 2004; Vaughan & Corkhill, 2017) or their toxicity (e.g., arsenopyrite is an extremely toxic pollutant; Vaughan & Corkhill, 2017). Sulfides generally contain pyramidal TS_3 groups in their structures (Fig. 1.3).

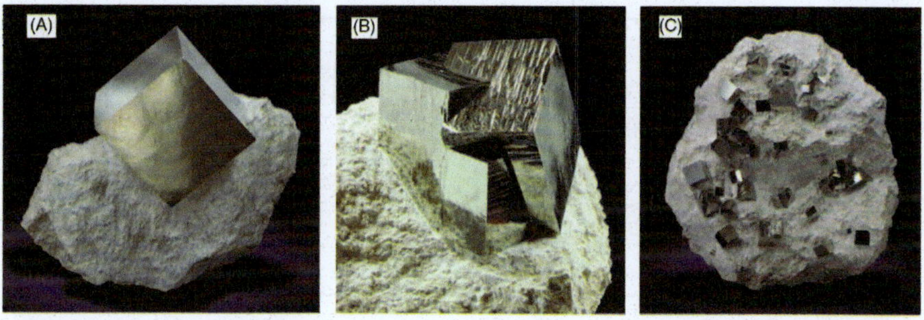

Figure 1.4 Pyrite crystals on a marlstone from Navajún, Spain: (A) pyrite crystal on 6 cm on edge on matrix, (B) pyrite crystal cluster on matrix with the growth hillocks on the cube face (largest cube is 5.5 cm on edge), (C) a 35-cm specimen of pyrite crystals on matrix. *(From Greene, E. S. (2015). Collecting Pyrite Crystals at Navajún, Spain. Rocks & Minerals, 90(1), 24–33. https://doi.org/10.1080/00357529.2015.973205).*

1.3.1 Pyrite

Pyrite is a brass-yellow mineral with a bright metal/lic luster that has a cubic habit (Fig. 1.4), which has a chemical composition of iron sulfide (FeS_2) and is the most common sulfide mineral. Pyrite forms at a wide range of temperature and occurs in nearly all rock types (igneous, hydrothermal, metamorphic sedimentary) and geological environments, often as microscopic crystals formed by both biological and geological processes (Butler et al., 2004; Madinabeitia et al., 2025). Pyrite is so common that many geologists would consider it to be a ubiquitous mineral (Wacey et al., 2011), but the factors that control its formation are still equivocal (Benning et al., 2000).

As per the Pourbaix (pE-pH) diagrams that show the thermodynamically stable phases of aqueous electrochemical systems, pyrite is supposed to be the mineral formed at low-temperature environments when the system is in equilibrium (Benning et al., 2000) as shown in Fig. 1.5. However, in reality, iron monosulfide is present (as a precursor mineral) prior to pyrite formation. The importance of an iron monosulfide precursor to pyrite formation is mackinawite Fe_9S_8—it has a chemical composition varying from $Fe_{0.87}S$ to $FeS_{1.1}$. Anoxic sediments indicate the presence and persistence of precursor iron monosulfides (acid-volatile sulfides, mackinawite and greigite) as key monosulfides in pyrite formation processes (Benning et al., 2000). When the state of oxidations of aqueous sulfur species is intermediate between sulfate and sulfide, pyrite formation is favored (Benning & Barnes, 1998).

Some studies of marine sediments, however, show that in most sediments, iron monosulfide phases and pyrite coexist (Schoonen, 2004). Thus the

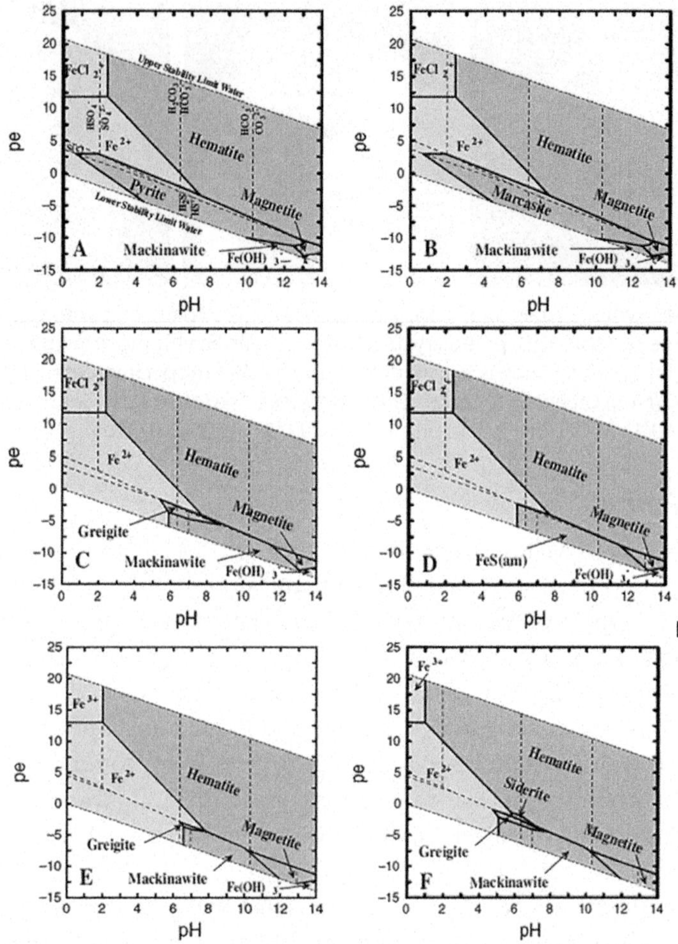

Figure 1.5 pE-pH diagrams illustrating the thermodynamic limits on sedimentary iron sulfides (25°C, 1 bar). (A) Equilibrium diagram for iron sulfides in seawater (see water composition from Parkhurst (Parkhurst & Appelo, 1999)), iron activity 10^{-6}, sulfur activity $10^{-2.551}$, C(IV) activity $10^{-3.001}$, troilite and pyrrhotite suppressed. (B) Same as previous panel, but pyrite, troilite and pyrrhotite suppressed. (C) Same as previous panel, but marcasite suppressed in addition. (D) Same as previous panel, but greigite and mackinawite suppressed in addition. (E) Same as (C) but solution composition changed from seawater to world average river water (Berner and Berner, 1996), with iron activity 10^{-6}, sulfur activity $10^{-3.902}$, C(IV) activity $10^{-3.06}$. F. Same as C but iron activity 10^{-6} increased to 10^{-3} and C(IV) increased to $10^{-2.5}$. All diagrams calculated using Geochemist's Work Bench™ (Bethke, 2002). In all diagrams, the thin, dashed lines define the speciation of the sulfur and C(VI) species (i.e., redox equilibration for carbon suppressed). For clarity, these lines are only labeled in panel (A). *(From Schoonen (2004)).*

pathways for pyrite formation remain debated. Experiments show that mackinawite, an iron monosulfide, is stable in reduced sulfur solutions at low temperatures for up to 4 months (Benning et al., 2000). According to Benning et al. (2000), pyrite forms negligibly in solutions with only aqueous H_2S below 100°C but can form in more oxidized conditions. They further state that oxidation experiments (pH 3.3–12) show that mackinawite transforms to pyrite in mildly oxidizing environments via a multi-step process involving sulfur species changes and the intermediate greigite. Oxidized surfaces of precursors or pyrite seeds accelerate this transformation.

1.3.1.1 Sedimentary pyrite

Sedimentary pyrite is iron sulfide that forms within sediments and sedimentary rocks, primarily through microbial activity in oxygen-poor (anoxic) environments, where SRB convert sulfate into sulfide, which then reacts with iron to create pyrite crystals, often appearing as microscopic cubes or spherical clusters called framboids (Berner, 1984; Raiswell & Berner, 1986). The formation of sedimentary pyrite explained here is in accordance with Schoonen (2004) where the role of bacteria, framboid formation and incorporation of impurities in the formation of pyrite have been proposed. Four key parameters are necessary: (1) metastable iron monosulfides coexisting with pyrite and appear to form first in anoxic sediments; (2) upon burial, the acid volatile sulfide (AVS)-to-pyrite ratio typically decreases; (3) pyrite is often present in the framboidal texture; and (4) marcasite is the dominant form of FeS_2 below a pH of ~5.

1.3.1.2 Framboid formation

Framboid formation involves the nucleation of microcrystals, often of iron monosulfide (such as mackinawite), followed by aggregation and growth into spheroidal shapes through processes such as magnetic properties of intermediate greigite or rapid nucleation in supersaturated solutions. These microaggregates are then transformed into more stable pyrite (FeS_2) through a replacement process. The resulting uniform size and packing of the microcrystals within a framboid are key to its characteristic structure (Butler & Rickard, 2000; Wilkin et al., 1996; Wilkin & Barnes, 1997). Sizes of the framboids are time-dependent and can be used to probe the evolution of the sedimentary environment at exceptionally high precision, diurnal scales (Rickard, 2019). On average, sedimentary pyrite framboids (~6 μm in size) take around 5 days to form, while syngenetic framboids ~6 μm in size forms in 3 days on average (Rickard, 2019). The largest known framboid measuring 250 μm took bout 3 years to form (Sweeney & Kaplan, 1973)

1.3.1.3 Incorporation of pyrite

Impurities such as Ni, As, Co and Cu are incorporated into sedimentary pyrite during formation, influencing its chemical composition and properties (Savage et al., 2008). This incorporation affects pyrite's electrochemical behavior and its ability to act as a geological indicator for paleo-redox conditions, with some impurities being desirable and others undesirable for specific applications such as mining (Lehner et al., 2007). The type, amount and position of these impurities depend on the local geochemical environment and can alter pyrite's texture and its role in regulating the biogeochemical cycles of these elements (Abraitis et al., 2004; Baya et al., 2021).

1.3.2 Magmatic sulfides

Magmatic sulfides form as the results of droplets of an immiscible sulfide-oxide liquid (Keays & Lightfoot, 2010), developing within silicate magma. Their origin can be a mantle or a crust. Certain elements, notably the group VIII transition metals Fe, Co, Ni, Pd, Pt, Rh, Ru, Ir, and Os, together with Cu and Au, partition strongly into sulfide oxide liquid and get concentrated—resulting in what is called a sulfide deposit (Naldrett, 1992).

1.3.2.1 Magmatic pyrite

Pyrite is a common sulfide mineral in many magmatic sulfide deposits (Naldrett, 2013), with pyrite abundance ranging from trace to being the predominant sulfide species (Steadman et al., 2021). The origin of pyrite in magmatic sulfide deposits is not always clear as pyrite can form as part of the primary base metal sulfide assemblage or as a secondary base metal sulfide (Mansur et al., 2021). Several studies have suggested that pyrite exsolved from monosulfide solid solution is enriched in IPGE relative to co-existing Po and Pn, whereas secondary Py has similar concentrations in IPGE relative to co-existing Po and Pn. Although this appears to be a global observation, it is not always the case (Mansur et al., 2021).

1.3.2.2 Magmatic pyrrhotite

Pyrrhotite is one of the most abundant iron sulfide minerals. It is found in nature to be commonly associated with pentlandite, quartz, ankerite

($CaFe(CO_3)_2$), pyrite, chalcopyrite, and other sulfide minerals (Miller et al., 2005). The pyrrhotite group of minerals is among the most common sulfide minerals (Wang, 2007). Pyrrhotite is routinely found in association with other commercial metallic compounds containing Ni, Zn, Au, etc. (Mansur et al., 2021). These minerals include troilite, monoclinic, and hexagonal pyrrhotite (Wang, 2007). They are normally considered a group, because the crystal structures of these minerals are derived from an NiAs-type structure (Simon & Ripley, 2011). Naturally occurring pyrrhotite group minerals are always associated with other sulfide minerals (Naldrett, 2013), such as pentlandite, violarite, pyrite and chalcopyrite. For example, pentlandite and violarite can exsolve from nickeliferous pyrrhotite in a range of temperatures (Wang, 2007). The rate of pyrrhotite oxidation is affected by factors like pH, temperature and the presence of microbes, with some research indicating it can dissolve faster than pyrite under certain conditions (Belzile et al., 2004; Moncur et al., 2009).

1.3.2.3 Magmatic chalcopyrite

Chalcopyrite is derived from the Greek words "chalkos," copper, and "pyrites" strike fire, and it is also known as copper pyrite (Baba et al., 2012). It is a brassy to golden yellow–colored mineral (Naldrett, 2013). Chalcopyrite ($CuFeS_2$) is the most common copper-bearing mineral on earth (Elliot & Watling, 2011). Chalcopyrite is of primary importance and occurs in igneous and metamorphic rock and in metalliferous veins (Baba et al., 2012). It contains many minerals including Cu, Zn, S and Fe. Chalcopyrite is the primary mineral that, by alteration and successive enrichment with Cu, produces Cu_5FeS, CuS, Cu_2S, and ending rarely as native Cu (Baba et al., 2012).

1.3.2.4 Arsenopyrite

In the context of AMD, magmatic arsenopyrite is a source of arsenic that oxidizes and dissolves in acidic, oxygenated water. Its oxidation releases soluble arsenic and other metals into the environment, creating toxic and acidic wastewater that pollutes surrounding areas through a complex process—dissolved arsenic (III) influences arsenopyrite oxidation rates and can lead to the formation of passivation layers on the mineral, which can suppress further dissolution and AMD production (Liu et al., 2025; Yadollahi et al., 2021).

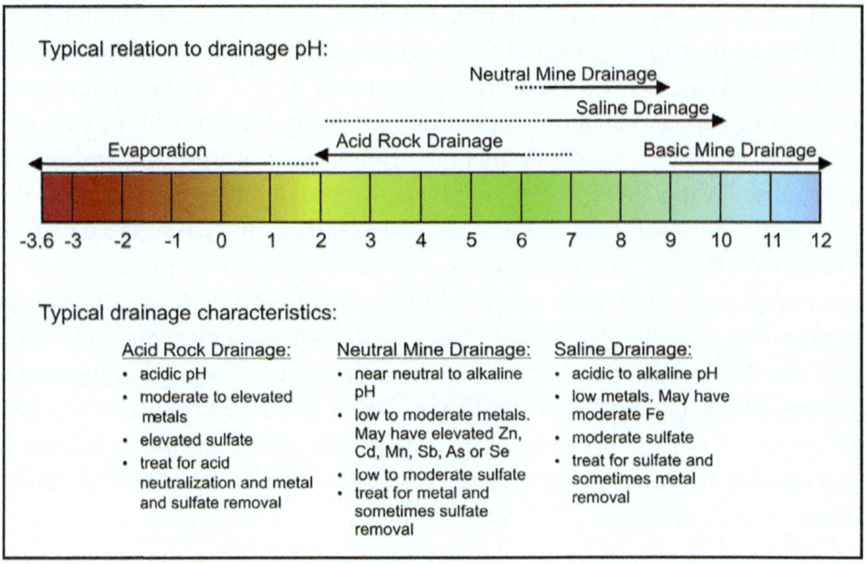

Figure 1.6 Acid, alkaline, and neutral mine drainage. *(From Nordstrom et al. (2015)).*

1.4 Rock types, ore types and AMD

The relationship between rock / ore types (sulfides) and AMD as well as their formation in various geological environments is given to enable the reader to have an overview of potential areas to find sulfides and hence AMD (Table 1.3).

1.5 Neutral, alkaline and saline drainage

In addition to acidic leachate or AMD, basic and circumneutral drainage containing high levels of contaminants can develop when overburden, tailings and exposed mined areas, such as shafts of abandoned or active mines, undergo weathering (Craw et al., 2023; Iwanoff, 1998; Masindi et al., 2022). Thus it is important to also consider other types of mining drainage, including neutral, alkaline and saline drainage. These types of mining drainages are presented in Fig. 1.6 by Nordstrom et al. (2015); in terms of pH range, mining drainages can be acidic pH <6, neutral (pH 6–9) and alkaline (pH >9).

Table 1.3 Major types of sulfide deposits.

Type	Major ore minerals	Metals extracted	Examples
Ores related to mafic and ultramafic intrusions			
Sulfide Ni deposits	po, pn, py, cpy, viol	Ni, Cu, PGM	Sudbury, Ontario, Canada
Merensky red Pt	Po, pn, cpy	Ni, Cu, PGM	Merensky reef, RSA
Ores related to felsic intrusive rocks			
Sn and W skarns	py, cass, sSph, cpy, wf	Sn, W	Pine creek, CA, USA
Zn–Pb skarns	py, Ssph, gn	Zn, Pb	Ban Ban, Australia
Cu skarns	py, cpy	Cu, Au	Carr Fork, Utah, USA
Porphyry Cu/Mo	py, cpy, bn, mbd,	Cu, Mo, Au	Bingham Canyon, Utah, climax
Polymetallic veins	py, cpy, gn, sph, ttd	Cu, Pb, Zn, Ag	Camsell River, NWT, Canada
High sulfidation ores	py, enar, cov, ten, Au	Cu, Au, Ag	Summitville, Co, USA
Ores related to marine mafic extrusive rocks			
Cyprus-type massive sulfides	py, cpy	Cu	Cyprus
Besshi-type massive sulfides	py, cpy, sph, gn	Cu, Pb, Zn	Japan
Ores related to subaerial felsic to mafic extrusive rocks			
Creede-type-epithermal veins	py, sph, gn, cpy, ttd, asp	Cu, pb, Zn, Ag, Au	Creede, CO, USA
Ores related to marine felsic mafic extrusive rocks			
Kuroko-type	py, sph, gn, cpy, ttd, asp	Cu, pb, Zn, Ag, Au,	Japan
Ores in clastic sedimentary rocks			
Quartz pebble U-Au	py,uran,gold	Au, U	Witwatersrand, RSA
Sandstone-hosted Pb-Zn	py, sph, gn	Pb, Zn, Cd	Laisvall, Sweden
Sedimentary exhalative Pb-Zn	py, sph, gn, cpy, ttd, po, asp	Cu, Pb, Zn, Au, Ag	Sullivan, BC, Canada, Ireland
Ores in carbonate rocks			
Mississippi Valley type	py, gn, sph	Zn, Pb, Cd, Ga, Ge	Missouri, USA

From Craig and Vaughan (1990)

asp, arsenopyrite; *Au*, gold; *bn*, bornite; *cass*, cassiterite; *cov*, covellite; *cpy*, chalcopyrite; *enar*, enargite; *gn*, galena; *mbd*, molybdenite; *PGM*, platinum-group minerals; *pn*, pentlandite; *po*, pyrrhotite; *py*, pyrite; *sph*, sphalerite; *ten*, tennantite; *ttd*, tetrahedrite; *uran*, uraninite; *viol*, violarite; *wf*, wolframite.

Table 1.4 Examples contaminants in alkaline mine drainage.

Constituent (mg/L except where noted)	Underground CMD, Vryheid Formation, South Africa	Stripa mine water, Sweden	Mine water from Lovozero Massif
T, C	n.d.	8.6	n.d.
pH	9.03	10.06	11.74
Specific conductance, lS/cm	2070	1090	n.d.
Constituent SO_4	376	57	n.d.
Cl	196	460	75
F	0.34	5.3	11,000
HCO_3	311	18	2666
SiO_2	2.52	18	9000–13,000
Na	235	218	24,038
K	26.4	0.44	222
Ca	71.2	94	n.d.
Mg	74.6	0.03	100
FeII	<0.001	0.0007	n.d.
Fe(T)	0.437	0.004	n.d.
Al	0.022	0.024	n.d.
Mn	0.003	<0.001	n.d.
Cu	<0.001	0.007	0.283
Pb	n.d.	<0.01	0.032
Zn	0.001	<0.002	n.d.
Cd	n.d.	n.d.	n.d.
Ni	<0.001	<0.001	n.d.
Co	<0.001	<0.005	n.d.
Cr	<0.001	<0.001	n.d.

From Nordstrom and Ptacek (2015).

1.5.1 Alkaline mine drainage

Mine waters may also be strongly alkaline (Craw et al., 2023; Ezekwe et al., 2013; Jang & Kwon, 2011; Shangguan et al., 2025; Skousen et al., 2017) with reported pH values exceeding 9 under specific hydrogeochemical settings (Nordstrom et al., 2015). Examples of contaminants associated with alkaline mine drainage are presented in Table 1.4.

1.5.2 Neutral mine drainage

Neutral mine drainage (NMD) is water contaminated by mining activities that is near-neutral pH (i.e., pH = 7) but contains high levels of dissolved metals and sulfates (e.g., Vriens et al., 2019). Alternative terminologies used to refer to NMD are contaminated neutral drainage (CND), neutral drainage (ND) (Plante et al., 2011a), neutral-rock drainage (NRD) (Vriens et al., 2019) or metal leaching (Vriens et al., 2020). Certain metals, such as Ni, Zn, Co, As and Sb, are soluble at near-neutral pH levels and can potentially contaminate mine effluents, even in the absence of acidic conditions (Plante et al., 2011b; Vriens et al., 2019). Oxyanionic elements such as As, Se and Mo exhibit elevated mobility at near-neutral pH (Vriens et al., 2020). The NMD has been reported in several locations worldwide, including the Force Crag mine in the United Kingdom, characterized by circumneutral pH and high Zn concentrations (Jarvis et al., 2023); the Erdouz Pb–Zn mines in Morocco (Goumih et al., 2022); arsenic-rich neutral drainage from gold mine wastes in Canada (Mehdaoui et al., 2025; Pakostova et al., 2023); Ni-rich neutral drainage from waste rocks at the Tio Hematite-Ilmenite deposit in Québec, Canada (Plante et al., 2010); contaminated neutral drainage in the Luanshya Copper belt in Zambia (Sracek et al., 2018) and Ni-rich effluent from the Hitura Ni mine in Finland (Heikkinen & Räisänen, 2009; Heikkinen et al., 2009). Examples of concentration of some elements detected in NRD are shown in Table 1.5.

Like for AMD, the leading cause of NMD is thought to be sulfide oxidation. In rocks with low sulfide content, H_2SO_4 initially forms from sulfide oxidation but is later neutralized by carbonate or silicate minerals. These processes—sulfide oxidation and neutralization—produce metal ions in solution, but they are immediately precipitated as oxyhydroxides, coprecipitated or adsorbed onto mineral surfaces (Bao et al., 2022; Gunsinger et al., 2006; Hunt et al., 1999). As a result, these ions do not appear in leachate solutions during kinetic tests. This is called the delayed onset of NMD (Demers et al., 2013). However, over time, as rocks weather, adsorbed metals can be desorbed or dissolved into water, leading to higher metal levels in the future. Consequently, NMD from waste rock can result from (1) sulfide oxidation when there is enough acid-buffering, (2) weathering of nonacid generating minerals in low-sulfide rocks, (3) dissolution of salts and (4) inadequate treatment of AMD, where pH is raised to near-neutral but some contaminants stay at high concentrations. Additionally, in reclaimed AMD wastes, the rate of acid production may decline to levels buffered by

Table 1.5 Examples of neutral-rock drainage chemical compositions from different mine sites around the world.

	Antamina, Peru (Vriens et al., 2019)	Hitura, Finland (Heikkinen & Räisänen, 2008; Heikkinen et al., 2009)	Lac Tio, Canada (Plante et al., 2014)	Beaver Brook, Canada (Radková, Jamieson, & Campbell, 2020)	Greens Creek, United States (Lindsay et al., 2009)	Giant Mine, Canada (Nordstrom, 2013)
pH	6.5–8.5	6.1–7.0	6.5–7.5	5.7–8.6	6.5–8.5	6.7
Ni (mg/L)	N/R	0.2–14.3	0.1–8.8	N/R	0–1	0.029
Zn (mg/L)	0.1–80	25–660	N/R	N/R	0–150	0.027
Mn (mg/L)	0.001–0.2	4.7–8.9	N/R	N/R	0–35	0.446
Co (mg/L)	N/R	0.05–7.2	N/R	N/R	N/R	<0.007
As (mg/L)	0.001–1.0	N/R	N/R	0–2.3	0–0.03	4060
Se (mg/L)	0.001–0.2	N/R	N/R	N/R	0–0.02	<0.03
Sb (mg/L)	0.001–0.2	N/R	N/R	0–26	0–0.06	11.9
Mo (mg/L)	0.0–1.0	N/R	N/R	N/R	0–0.02	0.07
SO_4^{2-} (g/L)	0.1–2	2.1–5.2	0.1–3.5	0.075–0.9	2–8	0.5
Ca (mg/L)	50–600	200–450	10–70	9–231	400–800	313

Adapted from Vriens et al. (2020).
NR, not reported.

neutralizing minerals, yet still cause metal leaching to occur. Thus, even if wastes are adequately treated, ongoing monitoring of mining wastes and the surrounding environments is necessary because NMD may develop many years later, rather than immediately, and might not be detected during early kinetic leaching tests.

1.5.3 Saline mine drainage

Salinity refers to the concentration of dissolved inorganic salts (e.g., NaCl) in water (Oseyemi et al., 2025; Qian et al., 2018). Saline mine drainage can be either alkaline, acidic or neutral. When the concentration of total dissolved solids (TDS) exceeds 10,000 mg/L, mine drainage is no longer AMD but saline mine drainage (SMD) (Ben Ali et al., 2019; Edmunds et al., 1987). The SMD is characterized by high levels of mineral ions such as Na^{+}, Cl^{-}, Mg^{2+}, Ca^{2+}, Ka^{+}, heavy metals and sulfates, which is a result of acid generation and neutralization processes (Ben Ali et al., 2019). Thus products of sulfide oxidation and dissolution of acid-neutralization minerals, including carbonates and silicates, constitute the SMD. The SMD can be determined by measuring pH, Eh, TDS and major ion concentrations (Pantsar-Kallio & Manninen, 1997), using either ion chromatography, ICP-OES or titration (AMIRA International, 2002; Craw, Rufaut, & Pillai, 2023; Pantsar-Kallio & Manninen, 1997). The environmental impacts of SMD include soil salinization, degradation of agricultural productivity, reduced water quality for domestic and industrial use, toxicity to aquatic organisms and alteration of freshwater ecosystems due to osmotic stress and ion imbalance (e.g., Beltrán, 1999).

1.6 Hydrological and climatic influence on AMD

Hydrology involves the occurrence, distribution, movement and properties of the waters of the earth and their relationship with the environment (Emel, Plisinski, & Rogan, 2014; Kant, Meena, & Singh, 2025). Climatic factors (e.g., temperature, wind and humidity) and topography (elevation and slope) influence the hydrological cycle (precipitation, runoff, evaporation and transpiration) (Adeyeri, 2025; Birimbayeva et al., 2025; Gnann et al., 2025). The interaction of this moving water with mine waste drives geochemical processes, such as oxidation and leaching, influencing the quality of mining drainage water (Vriens et al., 2020). However, the properties of mine wastes have an overall impact on the propagation of these geochemical processes. Physical properties such as particle size variation in tailings and waste rock piles create aeration, which allows for the infiltration of water and oxygen ingress (Chai et al., 2026; Lehmann et al., 2026; Song & Yanful, 2011), while mineralogical composition dictates the type and rate of geochemical reaction (Mafra et al., 2020; Schippers, 2004)).

Physical processes that link hydrology and AMD are (1) water transport, (2) gas transport and (3) heat transport (Gerke et al., 1998; Emel, Plisinski, & Rogan, 2014; Vriens, Plante, Seigneur, & Jamieson, 2020). Water plays a vital role in overall process that lead to acid generation and transport of contaminants in mine wastes. As water from rainfall or snow infiltrate the mine wastes, various solutes including product of rock weathering are dissolved (e.g., Clow & Drever, 1996; Nayak et al., 2023). These solutes are transported in aqueous form through various parts of the mine waste to groundwater (Söderberg et al., 2019). Aqueous transport of solutes is movement of water-soluble substances through porous media like soil and aquifers (Chen, Zhu, Young, & Susfalk, 2009; Slater & Comas, 2008). It is governed by three processes: (1) advection, (2) diffusion and (3) dispersion (Çallı et al., 2025). Advection is defined as the mechanical transport of solutes along with the bulk movement of water, driven by pressure gradient (Corey & Auvermann, 2003; Phillips & Castro, 2013). Diffusive transport is governed by concentration gradient, while dispersion is driven by differential advection (velocity variation) and entropy (Carrera, Ayora, Saaltink, & Dentz, 2012; Güven, Molz, & Melville, 1986; Leharne, 2019).

Gas transport is an important driver of rock weathering (Goddéris, Schott, & Brantley, 2019; Moore, Lichtner, White, & Brantley, 2012). For example, in addition to the roles of O_2 in sulfide oxidation, CO_2 also accelerates weathering of rocks (Zaharescu et al., 2019). Carbon dioxide dissolves in rainwater to break down silicate or carbonate rocks, converting the carbon into stable bicarbonates as shown in (Eqs. 1.5 and 1.6) (Donnini, Marchesini, & Zucchini, 2020; Holden et al., 2024).

$$CaMg(CO_3)_2 + 2CO_2 + 2H_2O \rightleftharpoons Ca^{2+} + Mg^{2+} + 4HCO_3^- \quad (1.5)$$

$$Mg_2SiO_4 + 4CO_2 + 4H_2O \rightarrow 4HCO_3^- + 2Mg^{2+} + H_4SiO_4 \quad (1.6)$$

Another critical parameter in mineral reactivity is heat transport. Heat energy drives chemical reactions in both natural and engineering systems. For example, waste landfills are always hotter than the surrounding environment because of heat generated due to various biogeochemical reactions within the wastes, including decomposition of organic matter (Yeşiller, Hanson, Kopp, & Yee, 2016). Yeşiller et al. (2015) report that waste landfills located in different climatic regions across the world generate temperature up to 60°C–90°C. A rise in temperature enhances microbial activity (Romero-Mujalli et al., 2019; Štyriaková et al., 2012). For example, some iron- and sulfur-oxidizing bacteria are active between 60°C and 65°C (Norris et al., 2013).

The physical processes described above occur in an interconnected and coupled manner (Sui et al., 2022). Sulfide oxidation produces acidity that induces the dissolution of carbonate and silicate minerals and the precipitation of secondary minerals (Vriens et al., 2020). The presence of the ferric ion (Fe^{3+}) also a product of these reactions, catalyzes electron and proton transfer, consequently enhancing the rate of sulfide oxidation. Oxidation of sulfide is exothermic reaction, estimated to release about 1100–1600 kJ/mol (Rohwerder et al., 1998). The exothermic energy released induces water and gas transport, both of which propagate oxidation throughout the mine waste. Thermal convection triggers vertical water migration by induced changes in fluid density and viscosity (Clauser, 2009). Although oxygen solubility and temperature are inversely related (Geng & Duan, 2010), the decreased fluid viscosity at higher temperatures enhances oxygen transport via diffusion (Dong & Hung, 1985).

The processes described above are internal geochemical reactions that are controlled by external processes such as particle size distribution, temperature and gas concentration and pressure and infiltration of water from rainfall as show in Figure 1.7 (Vriens et al., 2020).

Fig. 1.4 schematically illustrates the different physical transport processes and geochemical reactions controlling waste-rock drainage quality and quantity and how they interact.

Vriens et al. (2020) suggest that AMD in mine waste can be treated by controlling the processes that accelerate sulfide oxidation like oxygen influx or those that decelerate sulfide oxidation including decreased microbial activity, pore size clogging and secondary mineral precipitation. The Vriens et al.'s (2020) recommendation is practical. For example, applying hydraulic properties to design covers that limit oxygen and water influx to mine waste is a proven technique (Borghetti Soares et al., 2009; Kalonji Kabambi et al., 2017; Peppas et al., 2000; Xiong et al., 2025).

It is therefore suggested that knowledge of various parameters (in a multidisciplinary way) that control the formation of AMD can enable one to modify the existing treatment methods or develop novel methods as well as AMD prediction methods. This is possible in this era of AI and machine learning. For example, the lag time to the onset of AMD or NMD cases,

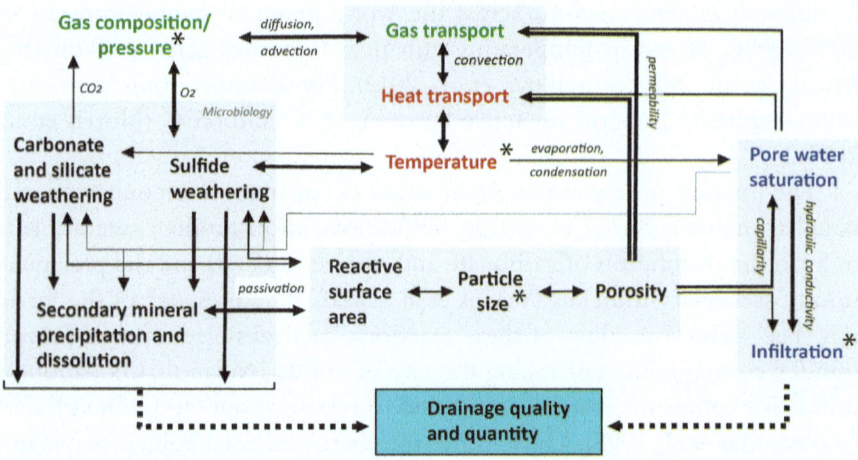

Figure 1.7 Interaction of physical processes with geochemical reactions and their influence in controlling waste rock drainage quality. Schematic illustration of major couplings and feedbacks between geochemical reactions and physical transport processes. Solid arrow lines indicate interactions or a control-hierarchy between different processes, with the width of the arrow being (qualitatively) proportional to the magnitude or strength of the feedback. External controls include that of site meteorology on infiltration, temperature, and gas pressure and that of site geology and communition method on waste-rock particle sizes, as indicated by the asterisks. *(From Vriens et al. (2020)).*

has not yet been found to have a solution, as identified in the next chapter. To bridge this gap, application of reactive transport models on leaching test results to study NMD or AMD is increasingly gaining popularity (Goddéris et al., 2019; Vriens et al., 2020; Zhang et al., 2024; El-Affani et al., 2025) . Detailed application of AI in AMD is discussed in Chapter 5.

1.7 Mobilization of potentially toxic substances in the environment by AMD

Acid mine drainage acts as a powerful geochemical catalyst, enhancing the generation, mobilization and toxicity of a wide range of pollutants (Table 1.6). Dissolution of toxic metal (loid)s such as As, Cd and Cr, at low pH is extensively documented (Ighalo et al., 2022; Jiao et al., 2023). This book puts particular emphasis on the interaction between AMD and commercial chemicals used in mineral extraction, specifically cyanide and ammonium nitrate–based explosives (Fortin et al., 2010; Zawadzka-Malota, 2015; Jaszczak et al., 2017; Hendry et al., 2018Marshall et al., 2020). The selection of these

Table 1.6 Impact of AMD on the mobility of potentially toxic substances.

Pollutant	Natural/geogenic source	Producing reaction	Anthropogenic/mining-related source	Environmental/health impact
Cyanide	Biogenic from plants and microbes (Azi et al., 2025; Xia et al., 2025); and geogenic processes such as emission volcanic activity (Curcic et al., 2025; Grimmett et al., 2025)	$NaCN + H^+ \rightarrow HCN + Na$	Extraction of gold and silver from their ores (Kuyucak & Akcil, 2013)	HCN is a very toxic gas. Inhalation of 100–300 ppm gaseous HCN results in death within 10–60 min, while 2000 ppm HCN causes death within 1 min. The LD50 for ingestion is 50,200 mg, or 1–3 mg/kg of body weight as HCN. For contact with unabraded skin, the LD50 is 100 mg/kg of body weight as HCN (Kuyucak & Akcil, 2013).
Nitrite nitrate	Biogenic from plants	Natural nitrification of ammonium in soils and regolith (e.g., Norton & Ouyang, 2019) $NH_4^+ + 2O_2 \rightarrow NO_3^- + H_2O + 2H^+$	Reaction of water with residual ammonium nitrate from explosives and gases such as NO_2 produced by explosives (Kidavu et al. 2025)	Acidification of surface water and soil (Wodrich et al., 2026) and eutrophication and brownfication due to elevated NO_3^- in water bodies (Jones et al., 2025; Singh et al., 2022)

(continued on next page)

Table 1.6 Impact of AMD on the mobility of potentially toxic substances—cont'd

Pollutant	Natural/geogenic source	Producing reaction	Anthropogenic/mining-related source	Environmental/health impact
Fluoride	Fluoropatite ($Ca_5(PO_4)_3F$), Bastnäsite: (REE)(CO_3)F, Fluorite or fluorspar (CaF_2) (Gysi & Williams-Jones, 2015; Rulis et al., 2004)	Weathering of fluoride-bearing minerals. For example,chemical weathering of fluorite in presence H_2SO_4 (from AMD) (Alves & Heubner, 2016) $CaF_2 + H_2SO_4 \rightarrow 2HF + CaSO_4$ or in presence of CO_2 and H_2O	Processing of Fluoride rich ores for REEs, Phosphate, etc. (Haneklaus et al., 2024).	Release of fluoride causing dental and skeletal fluorosis (Ali et al., 2019; Banyikwa, 2023; Hossain & Patra, 2020; Shaji et al., 2024; Vuhahula et al., 2009)

(continued on next page)

Phosphate	Apatite group minerals: $Ca_5(PO_4)_3$ (X) where X = F, OH, Cl. For example, Fluorapatite: $Ca_5(PO_4)_3F$, Hydroxyapatite: $Ca_5(PO_4)_3OH$, and Chloroapatite: $Ca_5(PO_4)_3Cl$ (Rulis et al. 2004) Octacalcium: phosphate (OCP), $Ca_8(HPO_4)_2(PO_4)_4.5H_2O$ (Idini & Frau, 2025)	Chemical weathering of phosphate minerals. For example, similar to wet process (Belboom et al., 2015), the AMD can enhance production of phosphoric acid from phosphate rich rocks $Ca_3(PO_4)_2 + 3H_2SO_4 \rightarrow 3CaSO_4 + 2H_3PO_4$	Processing of phosphate-rich ores for phosphate minerals and radioactive elements such as uranium and thorium (Michael et al., 2024)	Eutrophication of receiving waters (Oliveira & Machado, 2013); mobilization of Cd, U and REEs associated with phosphates (Haneklaus et al., 2024)
Organic acids	Peat, coal seams, carbonaceous shales	Microbial decomposition of organic matter (Adeleke et al., 2017)	Organic-rich tailings, reclaimed soils with organic amendments (Campbell et al., 2017; Wang et al., 2025)	Metal complexation increasing mobility and bioavailability of Cu, Ni, Zn, Pb (Kim et al., 2013; 2008; Yao et al., 2022) and dissolved oxygen depletion in waters (Goyne et al., 2006)

AMD, acid mine drainage; *REE*, rare earth element.

two pollutants is based on (1) their acute toxicity, which is prone to triggering social conflicts among farmers, pastoralists and miners—especially within the small-scale mining sector where these economic activities commonly intersect; and (2) their diverse sources, ranging from geogenic, biogenic to anthropogenic origins, which, if not properly identified may mislead justice processes concerning alleged pollution.

1.7.1 Cyanide

Hydrocyanic acid or hydrogen cyanide (HCN) gas is a weak acid ($pKa = 9.22, 25°C$) and pH sensitive (Petrikovics et al., 2015). In waters with pH below 8.5, HCN exists as gas whereas at a pH of 10.5 or higher, it exists as a free cyanide anion (CN) (Kuyucak & Akcil, 2013). Hydrocyanic acid gas is very toxic when ingested, inhaled, or absorbed via skin (Alan-Albayrak & Simonsen, 2025). For instance, inhalation of 100–300 ppm gaseous HCN results in death within 10–60 minutes, while 2000 ppm HCN causes death within 1 minute (Kuyucak & Akcil, 2013). The rapid death caused by HCN is due to asphyxiation (lack of oxygen) or histotoxic hypoxia (cells cannot use oxygen), which leads to cardiac arrest (Egekeze & Oehme, 1980; Fortin et al., 2010). Cyanides inhibit the final stage of adenosine triphosphate (ATP) synthesis in the mitochondrial electron transport chain, the oxidative phosphorylation which renders the body unable to generate energy in the form of ATP from oxygen (Abdollahi & Hosseini, 2014). Due to its high affinity for iron, cyanide binds to the heme a_3-Cu_B binuclear center (the active cite for O_2) within the cytochrome c oxidase enzyme, inhibiting the transfer of electrons to their final acceptor molecule, O_2 (Zuhra & Szabo, 2022). In eukaryotes, metabolic energy in the form of ATP is generated through metabolic reactions. These processes, which involve electron transfer (redox) reactions, channel energy into high-energy, electron-rich carrier molecules, namely nicotinamide adenine dinucleotide (NADH) and Flavin adenine dinucleotide ($FADH_2$) (Osellame et al., 2012). These reduced cofactors donate electrons to the mitochondrial electron transport chain (ETC), as illustrated in Fig. 1.8A. CIV is cytochrome c oxidase, or simply complex IV. It consists of four redox centers comprising bimetallic CuA, heme a and heme a_3-Cu_B (the oxygen-binding catalytic site). Electrons are first accepted by Cu_A, the reduced Cu_A donates electrons to heme a, which transfers them to the heme a_3-Cu_B binuclear redox center (Zuhra & Szabo, 2022). The a_3-Cu_B binuclear redox center is a catalytic site where oxygen is reduced to H_2O and H^+ is subsequently released. The heme

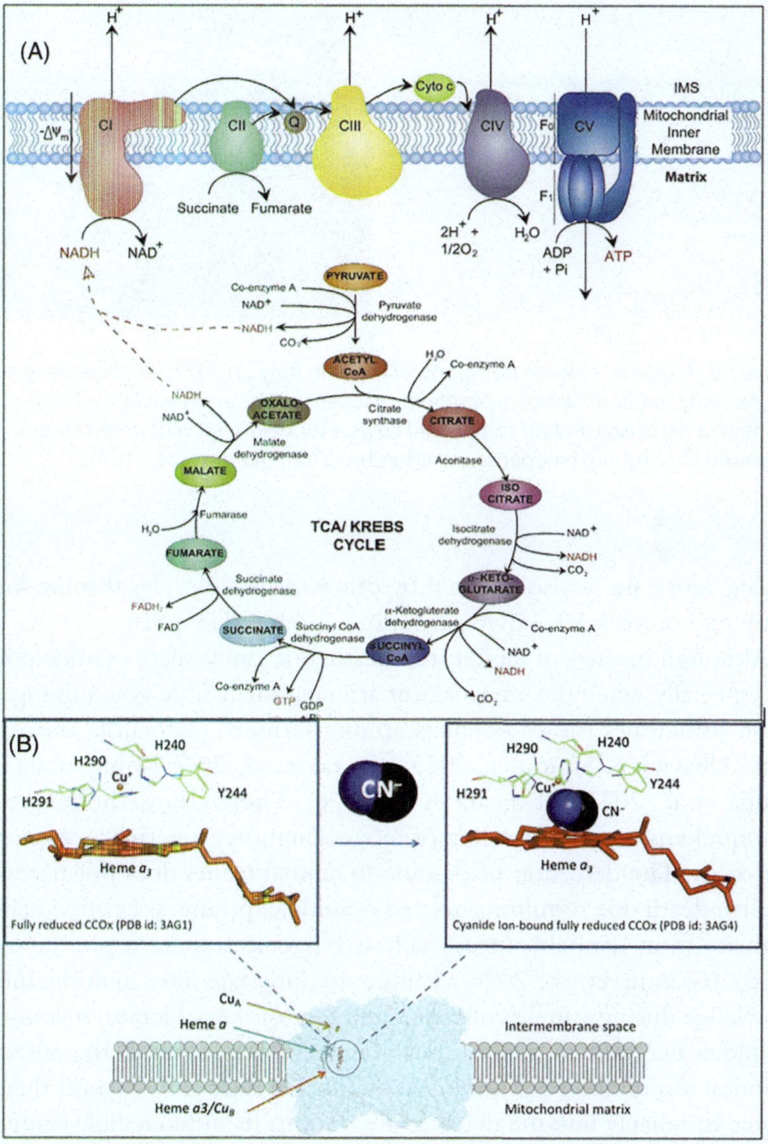

Figure 1.8 Effect of cyanide on electron transport system of eukaryotes: (A) Normal bioenergetics of the electron transport chain and the TCA/Kerbs cycle which involves conversion of catabolic product of carbohydrate (pyruvate) to high-energy molecules (NADH, GTP, and $FADH_2$) and to oxidative phosphorylation of ADP to form ATP (Osellame et al., 2012). (B) cyanide binds cytochrome c oxidase (CCOx) at heme a_3-Cu_B inhibiting the O_2 uptake (Osellame et al., 2012). *((A) From Osellame et al. (2012); (B) From Zuhra et al. (2022)).*

(A) (B)

Figure 1.9 Biogenic cyanide poisoning of cattle in Italy on 2022: (A) some dead cows after feeding *Sorghum bicolor* × *Sorghum sudanense* cultivar grown in the field, Sommariva del Bosco and (B) some dead cows after being fed with green chop composed of Sorghum halepense in—Moretta. *(From Giantin et al. (2024)).*

a_3-Cu_B active site is also targeted by other small molecules that are known to inhibit complex VI activity, including cyanide (Fig. 1.8B).

Although in cases of rapid cattle death, one can suspect cyanide poisoning, especially when the scene is near artisanal small-scale gold mining areas, which sometimes causes conflicts among farmers, pastoralists and miners (Alan-Albayrak & Simonsen, 2025; Jaszczak et al., 2017; Kohio et al., 2025; Meutia et al., 2023; Mhangara et al., 2026). These claims, though, require a comprehensive understanding of biogeochemistry, veterinary science and toxicology. The detection of cyanide in animal tissues does not necessarily confirm death due to mining-related cyanide exposure, as cyanide may also originate from natural sources, such as biogenic sources (e.g., cyanogenic plants) (Giantin et al., 2024). Failure to integrate this multidisciplinary knowledge during sample collection and forensic toxicological investigation of sudden livestock mortality in mining areas may result in the false attribution of responsibility to mining activities. Forensic analysts must therefore be able to reliably link the detected toxicant to its most credible source.

An example of biogenic cyanide poisoning is described by Giantin et al. (2024). In their work entitled "*Overview of Cyanide Poisoning in Cattle from Sorghum halepense and S. bicolor Cultivars in Northwest Italy*," Giantin et al. (2024) report that 46/160 cows died within 20–30 minutes of eating cultivar Suzy, a hybrid of *Sorghum bicolor* and *Sorghum sudanense* (Fig. 1.9A), while 5/20 lactating cows died immediately after being fed green chop composed of Sorghum halepense (Fig. 1.9B). Different Sorghum cultivars may accumulate dhurrin (cyanogenic glycoside) concentrations that can trigger lethal poisoning in cows, particularly if animals feed on young plants or sprouts under adverse climatic conditions (drought, tropical temperatures) (Giantin et al., 2024). Some plants, such as sorghum, produce dhurrin in summer as a defense mechanism against herbivores and potentially as a nitrogen reserve during periods of environmental stress, specifically drought (Kumar et al., 2025).

Giantin et al. (2024) shows that rapid death of cattle can be caused by naturally occurring cyanide in plants. The HCN occurs naturally through biogenic processes by certain microbes, such as *Pseudomonas brassicacearum* (Kumar et al., 2025) and some plants, for example, Cassava and sorghum (Azi et al., 2025; Xia et al., 2025); and geogenic processes such as volcanic activity emission (Curcic et al., 2025; Grimmett et al., 2025). The anthropogenic sources of HCN are burning of fossil fuel (Dagaut et al., 2008), cyanidation extraction of gold and silver (Kuyucak & Akcil, 2013) and fires involving wools, silks and synthetic nitrogen-containing polymers (e.g., urethane and nylon) (Jasani, 2015). Biogenic cyanide is highly toxic to both human and ruminants. Ruminants are susceptible to poisoning by cyanogenic plants because of the more efficient hydrolysis of glycosides by ruminal enzymes. In ruminants, the pH of the rumen is neither strongly acidic nor alkaline and it comprises a large flora of microorganisms with substantial amounts of enzymes.

1.7.2 Nitrate/Nitrite

Like cyanide, nitrate/nitrite can cause rapid deaths in cattle. The mechanism involves the conversion of the red, oxygen-carrying protein hemoglobin into methemoglobin, a brown protein that cannot transport oxygen (Hall, 2018; Pjetraj et al., 2025; Wang et al., 2025). After cattle ingest a high amount of nitrate from contaminated plants or water, microorganisms in the rumen rapidly convert it to nitrite. The nitrite is absorbed into the blood, where it oxidizes the iron in hemoglobin, forming methemoglobin, which cannot transport oxygen. The blood loses its oxygen-carrying capacity. Consequently, tissues and vital organs are starved of oxygen, leading to severe hypoxia (lack of oxygen) and death (Tulisiak, Loesing, Ljungberg, & Kaczor, 2025; Pjetraj et al., 2025).

The source of nitrate is both natural and anthropogenic. Examples of natural sources include lightning, which provides heat energy for the reaction of atmospheric O_2 and N_2; microbial soil activity, such as nitrification, which oxidizes NH_4^+ to NO_3^- (Subba Rao et al., 2017); volcanoes, which emit gases including SO_2 and NO_2 (Mather et al., 2004); and oceans, which emit N_2O (Craig & Gordon, 1963; Hashimoto et al., 1999). The anthropogenic sources of toxic nitrogen oxides are diverse, including the oxidation and hydration of nitrogen oxides from the combustion of fossil fuels and explosives (House & House, 2016), the reaction of water with residual ammonium nitrate from fertilizers used in agriculture, and the use of nitrate-based explosives in mine blasting (Kidavu, Chaudhary & Soni, 2025). Oruc et al. (2010) report that nine horses that died at a farm in Bursa, Turkey, between February and April 2008 were due to biogenic nitrate poisoning, as high nitrate concentrations were determined in the bodies of the horses as well as in some grass and alfalfa samples used to feed the animals.

In conclusion, although AMD is widely studied, adverse effects of mine effluents can occur across a broad range of pH, from acidic to alkaline. Thus research should move beyond acidic drainage of sulfide minerals by fully integrating neutral, alkaline and saline drainages into conceptual and risk-assessment frameworks. Priority should be given to resolving the geochemical controls on toxic elements' mobility and remobilization at nonacidic pH, including the role of secondary acids in enhancing mineral weathering and contaminant release. To distinguish geogenic, biogenic and anthropogenic contaminant sources is critical to reduce misinterpretation of environmental poisoning events suspected in mining areas.

References

Abdollahi, M., & Hosseini, A. (2014). *Cyanogen chloride encyclopedia of toxicology: Third ed.* (pp. 1096–1099). Elsevier Available from. http://www.sciencedirect.com/science/book/9780123864550 . doi:10.1016/B978-0-12-386454-3.00484-X.

Abraitis, P. K., Pattrick, R. A. D., & Vaughan, D. J. (2004). Variations in the compositional, textural and electrical properties of natural pyrite: A review. *International Journal of Mineral Processing, 74*(1–4), 41–59. doi:10.1016/j.minpro.2003.09.002.

Adeleke, R., Nwangburuka, C., & Oboirien, B. (2017). Origins, roles and fate of organic acids in soils: A review. *South African Journal of Botany, 108*, 393–406. doi:10.1016/j.sajb.2016.09.002.

Adeyeri, O. E. (2025). Hydrology and Climate Change in Africa: Contemporary Challenges, and Future Resilience Pathways. *Water (Switzerland), 17*(15), 1–20. https://doi.org/10.3390/w17152247.

Akinwekomi, V., Maree, J. P., Masindi, V., Zvinowanda, C., Osman, M. S., Foteinis, S., Mpenyana-Monyatsi, L., & Chatzisymeon, E. (2020). Beneficiation of acid mine drainage (AMD): A viable option for the synthesis of goethite, hematite, magnetite, and gypsum – Gearing towards a circular economy concept. *Minerals Engineering, 148*, 106204. doi:10.1016/j.mineng.2020.106204.

Alan-Albayrak, E., & Simonsen, U. (2025). Cyanide beyond toxicity: A systematic review of its effects on vascular function. *Basic and Clinical Pharmacology and Toxicology, 137*(5). Available from http://onlinelibrary.wiley.com/journal/10.1111/(ISSN)1742-7843 . doi:10.1111/bcpt.70124.

Ali, S., Fakhri, Y., Golbini, M., Kumar Thakur, S., Alinejad, A., Parseh, I., Shekhar, S., & Bhattacharya, P. (2019). Concentration of fluoride in groundwater of India: A systematic review, meta-analysis and risk assessment. *Groundwater for Sustainable Development, 9*, 100224. doi:10.1016/j.gsd.2019.100224.

Almås, A. R., & Manoko, M. L. K. (2012). Trace element concentrations in soil, sediments, and waters in the vicinity of Geita Gold Mines and North Mara Gold Mines in Northwest Tanzania. *Soil and Sediment Contamination, 21*(2), 135–159. doi:10.1080/15320383.2012.649372.

Alves, H., & Heubner, U. (2016). Aqueous corrosion of nickel and its alloys. *Reference module in materials science and materials engineering*: 1. Elsevier Ltd July 2025 https://doi.org/10.1016/b978-0-12-803581-8.10114-6 .

AMIRA International. (2002). *ARD test handbook: Prediction and kinetic control of acid mine drainage (Project P387A)*. Ian Wark Research Institute and Environmental Geochemistry Pty Ltd.

Anawar, H. Md (2015). Sustainable rehabilitation of mining waste and acid mine drainage using geochemistry, mine type, mineralogy, texture, ore extraction and climate knowledge. *Journal of Environmental Management, 158*, 111–121. doi:10.1016/j.jenvman.2015.04.045.

Azi, F., Li, Z., & Xu, P. (2025). Expanding Yarrowia lipolytica's metabolic potential for detoxification of cyanogenic glycosides in edible plants. *Communications Biology, 8*(1). doi:10.1038/s42003-025-07628-5.

Baba, A. A., Ayinla, K. I., Adekola, F. A., Ghosh, M. K., Ayanda, O. S., Bale, R. B., Sheik, A. R., & Pradhan, S. R. (2012). A review on novel techniques for chalcopyrite ore processing. *International Journal of Mining Engineering and Mineral Processing, 1*(1), 1–16. https://doi.org/10.5923/j.mining.20120101.01.

Bai, T., He, S., Li, Y., Deng, J., Sang, F., Zhao, Y., Wang, H., He, T., Zhang, K., Qiu, Y., Wang, P., Ye, C., Zhang, Y. i, & Wang, Y. i (2025). Unexpected suppressive fungal diversity and stimulative soil carbon loss under soil acidification in an alkaline grassland. *Functional Ecology, 39*(1), 114–127. doi:10.1111/1365-2435.14685.

Banyikwa, A. T (2023). Geochemistry and sources of fluoride and nitrate contamination of groundwater in six districts of the Dodoma region in Tanzania, *Scientific African* p. e01783. doi:10.1016/j.sciaf.2023.e01783.

Bao, Z., Bain, J., Saurette, E., Zou Finfrock, Y., Hu, Y., Ptacek, C. J., & Blowes, D. W. (2022). Mineralogy-dependent sulfide oxidation via polysulfide and thiosulfate pathways during weathering of mixed-sulfide bearing mine waste rock. *Geochimica et Cosmochimica Acta, 317*, 523–537. Available from http://www.journals.elsevier.com/geochimica-et-cosmochimica-acta/ . doi:10.1016/j.gca.2021.10.012.

Baya, C., Pape, P. L. E., Baptiste, B., Brest, J., Landrot, G., Elkaim, E., Noël, V., Blanchard, M., Ona-Nguema, G., Juillot, F., & Morin, G. (2021). Influence of trace level As or Ni on pyrite formation kinetics at low temperature. *Geochimica et Cosmochimica Acta, 300*, 333–353. doi:10.1016/j.gca.2021.01.042.

Benning, L. G., & Barnes, H. L. (1998). In situ Determination of the Stability of Iron Monosulphides and Kinetics of Pyrite Formation. *Mineralogical Magazine, 62A*(1), 151–152

Belboom, S., Szöcs, C., & Léonard, A. é (2015). Environmental impacts of phosphoric acid production using di-hemihydrate process: A Belgian case study. *Journal of Cleaner Production, 108*, 978–986. doi:10.1016/j.jclepro.2015.06.141.

Beltrán, J. M. (1999). Italy Irrigation with saline water: Benefits and environmental impact. *Agricultural Water Management, 40*, 183–194. doi:10.1016/S0378-3774(98)00120-6.

Belzile, N., Chen, Y. W., Cai, M. F., & Li, Y. (2004). A review on pyrrhotite oxidation. *Journal of Geochemical Exploration, 84*(2), 65–76. Available from http://www.sciencedirect.com/science/journal/03756742 . doi:10.1016/j.gexplo.2004.03.003.

Ben Ali, E. H., Neculita, C. M., Molson, J. W., Maqsoud, A., & Zagury, G. J. (2019). Efficiency of batch biochemical reactors for mine drainage treatment at low temperature and high salinity. *Applied Geochemistry, 103*(January), 40–49. https://doi.org/10.1016/j.apgeochem.2019.01.014.

Benning, L. G., Wilkin, R. T., & Barnes, H. L. (2000). United States Reaction pathways in the Fe-S system below 100°C. *Chemical Geology, 167*, 25–51. doi:10.1016/S0009-2541(99)00198-9.

Berner, R. A. (1984). Sedimentary pyrite formation: An update. *Geochimica et Cosmochimica Acta, 48*(4), 605–615. doi:10.1016/0016-7037(84)90089-9.

Bhattacharya, A., Routh, J., Jacks, G., Bhattacharya, P., & Mörth, M. (2006). Environmental assessment of abandoned mine tailings in Adak, Västerbotten district (northern Sweden). *Applied Geochemistry, 21*(10), 1760–1780. doi:10.1016/j.apgeochem.2006.06.011.

Birimbayeva, L., Makhmudova, L., Alimkulov, S., Tursunova, A., Tigkas, D., Abayev, N., Dostayeva, A., Birimbayev, Z., Alzhanov, O., Rodrigo-Clavero, M. E., & Rodrigo-Ilarri, J. (2025). Reaction of Minimal Streamflow to Natural Factors in the Context of Climate Uncertainty. *Water Resources Management,, 39*(12), 6009–6025. https://doi.org/10.1007/s11269-025-04238-y.

Borghetti Soares, A., Ubaldo, M. de O., de Souza, V. P., Soares, P. S. M., Barbosa, M. C., & Mendonça, R. M. G. (2009). Design of a Dry Cover Pilot Test for Acid Mine Drainage Abatement in Southern Brazil. I: Materials Characterization and Numerical Modeling. *Mine Water and the Environment, 28*(3), 219–231. https://doi.org/10.1007/s10230-009-0077-5

Bonnissel-Gissinger, P., Alnot, M., Ehrhardt, J. J., & Behra, P. (1998). Surface oxidation of pyrite as a function of pH. *Environmental Science and Technology, 32*(19), 2839–2845. doi:10.1021/es980213c.

Bulaev, A., Kadnikov, V., Elkina, Y., Beletsky, A., Artykova, A., Kolosoff, A., Ravin, N., & Mardanov, A. (2025). Transcriptomic analysis of the strain Acidiplasma sp. YE-1 During the oxidation of sulfide minerals pyrite and arsenopyrite. *International Journal of Molecular Sciences, 26*(19), 9287. doi:10.3390/ijms26199287.

Butler, I. B., Böttcher, M. E., Rickard, D., & Oldroyd, A. (2004). Sulfur isotope partitioning during experimental formation of pyrite via the polysulfide and hydrogen sulfide pathways: Implications for the interpretation of sedimentary and hydrothermal pyrite isotope records. *Earth and Planetary Science Letters, 228*(3-4), 495–509. doi:10.1016/j.epsl.2004.10.005.

Butler, I. B., & Rickard, D. (2000). Framboidal pyrite formation via the oxidation of iron (II) monosulfide by hydrogen sulphide. *Geochimica et Cosmochimica Acta, 64*(15), 2665–2672. doi:10.1016/S0016-7037(00)00387-2.

Çallı, K. Ö., Chiogna, G., Bittner, D., Sivelle, V., Labat, D., Richieri, B., … Hartmann, A. (2025). Karst Water Resources in a Changing World: Review of Solute Transport Modeling Approaches. *Reviews of Geophysics, 63*(1). https://doi.org/10.1029/2023RG000811.

Campbell, D., Stewart, K., Spiers, G., & Beckett, P. (2017). Growth and metal uptake of canola and sunflower along a thickness gradient of organic-rich covers over metal mine tailings. *Ecological Engineering, 109*, 133–139. doi:10.1016/j.ecoleng.2017.08.019.

Carrera, J., Ayora, C., Saaltink, M. W., & Dentz, M. (2012). Radionuclide migration: Coupling transport and chemistry. *Radionuclide Behaviour in the Natural Environment: Science, Implications and Lessons for the. Nuclear Industry*, 302–383. https://doi.org/10.1533/9780857097194.2.302.

Chai, C., Song, H., Feng, Z., Zou, Y., & Munkhbat, B. (2026). Improving water retention of loess and promoting plant growth in arid mining areas using cement-bonded coal gangue porous material. *Results in Engineering, 29*, 109209. https://doi.org/10.1016/j.rineng.2026.109209.

Chen, L., Zhu, J., Young, M. H., & Susfalk, R. B. (2009). An integrated approach for modeling solute transport in streams and canals with applications. *Journal of Hydrology, 378*(1–2), 128–136. https://doi.org/10.1016/j.jhydrol.2009.09.012.

Chen, W. B., Chen, H., Hsiao, S. C., Liang, T. Y., Shih, H. J., & Chang, C. H. (2026). A trend-based ecological indicator framework for spatially classifying ocean acidification risk to global coral reefs. *Ecological Indicators, 182*(1), 114545. https://doi.org/10.1016/j.ecolind.2025.114545.

Cheng, K. Y., Acuña, C. R., Kaksonen, A. H., Esslemont, G., & Douglas, G. B. (2021). Treatment of neutral gold mine drainage by sequential in situ hydrotalcite precipitation, and microbial sulfate and cyanide removal. *Science of the Total Environment, 801*. Available from www.elsevier.com/locate/scitotenv . doi:10.1016/j.scitotenv.2021.149613.

Chiriță, P., & Rimstidt, J. D. (2014). Pyrrhotite dissolution in acidic media. *Applied Geochemistry, 41*, 1–10. https://doi.org/10.1016/j.apgeochem.2013.11.013.

Clauser, C. (2009). Heat Transport Processes in the Earth's Crust. *Surveys in Geophysics, 30*(3), 163–191. https://doi.org/10.1007/s10712-009-9058-2.

Clow, D. W., & Drever, J. I. (1996). Weathering rates as a function of flow through an alpine soil. *Chemical Geology, 132*(1–4), 131–141. https://doi.org/10.1016/S0009-2541(96)00048-4.

Corey, A. T., & Auvermann, B. W. (2003). Transport by Advection and Diffusion Revisited. *Vadose Zone Journal, 2*(4), 655–663. https://doi.org/10.2136/vzj2003.6550.

Craig, H., & Gordon, L. I. (1963). Nitrous oxide in the ocean and the marine atmosphere. *Geochimica et Cosmochimica Acta, 27*(9), 949–955. doi:10.1016/0016-7037(63)90104-2.

Craig, J. R. & Vaughan, D. J. (1990). Compositional and textural variations of the major iron and base-metal sulphide minerals. In: Sulphide deposits—Their origin and processing (pp. 1–16). Springer Netherlands. https://doi.org/10.1007/978-94-009-0809-3_1.

Craw, D., Rufaut, C., & Pillai, D. (2023). Evolution of alkaline mine drainage and unique biodiversity on soil-free mine substrates, Southern New Zealand. *Mine Water and the Environment, 42*(1), 3–23. doi:10.1007/s10230-023-00913-x.

Curcic, M., Miljaković, E. A., & Antonijević, B. (2025). Cardiotoxicity of gaseous poisons. *Advances in Biochemistry in Health and Disease, 33*, 401–414. Available from https://www.springer.com/series/7064 . doi:10.1007/978-3-031-90598-8_21.

Dagaut, P., Glarborg, P., & Alzueta, M. (2008). The oxidation of hydrogen cyanide and related chemistry. *Progress in Energy and Combustion Science, 34*(1), 1–46. doi:10.1016/j.pecs.2007.02.004.

Dash, J., Ojha, R., & Pradhan, D. (2025). Progress in bioleaching and its mechanism: A short review. *Discover Environment, 3*(1), 238. doi:10.1007/s44274-025-00454-w.

Deberdt, R., Smith, N. M., Calderon, J. L., & McCall, S. K. (2025). Critical minerals lists for low-carbon transitions: Reviewing their structure, objectives, and limitations. *Energy Research and Social Science, 127*. Available from https://www.sciencedirect.com/science/journal/22146296 . doi:10.1016/j.erss.2025.104252.

Demers, I., Molson, J., Bussière, B., & Laflamme, D. (2013). Numerical modeling of contaminated neutral drainage from a waste-rock field test cell. *Applied Geochemistry, 33*, 346–356. doi:10.1016/j.apgeochem.2013.02.025.

Dold, B. (2017). Acid rock drainage prediction: A critical review. *Journal of Geochemical Exploration, 172*, 120–132. doi:10.1016/j.gexplo.2016.09.014.

Doman, A., Doka, E., Garai, D., Bogdandi, V., Balla, G., Balla, J., & Nagy, P. (2023). Interactions of reactive sulfur species with metalloproteins. *Redox Biology, 60*, 102617. https://doi.org/10.1016/j.redox.2023.102617.

Dominique, G., Dossa, K. F., Larivière, D., & Khasa, D. P. (2025). Environmental, health and social acceptability issues associated with rare earth elements: A systematic literature review. *Discover Environment, 3*(1). Available from https://link.springer.com/journal/44274. doi:10.1007/s44274-025-00274-y.

Dong, Y., & Hung, Y. (1985). Effect of temperature on oxygen transfer rate in wastewater treatment. *International Journal of Environmental Studies, 24*(2), 125–135. https://doi.org/10.1080/00207238508710186.

Donnini, M., Marchesini, I., & Zucchini, A. (2020). A new Alpine geo-lithological map (Alpine-Geo-LiM) and global carbon cycle implications. *Bulletin of the Geological Society of America, 132*(9–10), 2004–2022. https://doi.org/10.1130/B35236.1.

Duan, Y., Zhang, J., Petropoulos, E., Zhao, J., Jia, R., Wu, F., Chen, Y., Wang, L., Wang, X., Li, Y., & Li, Y. (2025). Soil acidification destabilizes terrestrial ecosystems via decoupling soil microbiome. *Global Change Biology, 31*(4). doi:10.1111/gcb.70174.

Duverger, A., Berg, J. S., Busigny, V., Guyot, F., Bernard, S., & Miot, J. (2020). Mechanisms of pyrite formation promoted by sulfate-reducing bacteria in pure culture. *Frontiers in Earth Science, 8*. Available from https://www.frontiersin.org/journals/earth-science . doi:10.3389/feart.2020.588310.

Edmunds, W. M., Cook, J. M., Darling, W. G., Kinniburgh, D. G., Miles, D. L., Bath, A. H., Morgan-Jones, M., & Andrews, J. N. (1987). Baseline geochemical conditions in the Chalk aquifer, Berkshire, U.K.: A basis for groundwater quality management. *Applied Geochemistry, 2*(3), 251–274. doi:10.1016/0883-2927(87)90042-4.

Egekeze, J. O., & Oehme, F. W. (1980). Cyanides and their toxicity: A literature review. *Tijdschrift voor diergeneeskunde, 105*(8).

Elghali, A., Benzaazoua, M., Taha, Y., Amar, H., Ait-khouia, Y., Bouzahzah, H., & Hakkou, R. (2023). Prediction of acid mine drainage: Where we are. *Earth-Science Reviews, 241*, 104421. doi:10.1016/j.earscirev.2023.104421.

El-Affani, L., Bussière, B., Qureshi, A., & Plante, B. (2025). Geochemical behavior of amended and non-amended mine tailings as cover materials for acid mine drainage control: Column tests and reactive transport modeling. *Journal of Contaminant Hydrology, 272*(April). https://doi.org/10.1016/j.jconhyd.2025.104564

Elliot, A. D., & Watling, H. R. (2011). Chalcopyrite formation through the metathesis of pyrrhotite with aqueous copper. *Geochimica et Cosmochimica Acta, 75*(8), 2103–2118. doi:10.1016/j.gca.2011.01.033.

Emel, J., Plisinski, J., & Rogan, J. (2014). Monitoring geomorphic and hydrologic change at mine sites using satellite imagery: The Geita Gold Mine in Tanzania. *Applied Geography, 54*, 243–249. https://doi.org/10.1016/j.apgeog.2014.07.009.

Eswar, I., Majumder, A., mohamed, S. HR., Logesh, K., alavudeen, S. MI. S, Hariprakash, M., & Sundarraj, D. K (2025). Seaweed responses to ocean acidification: Global impacts on growth, biochemical composition, and CO2 mitigation potential. *Discover Oceans, 2*(1). doi:10.1007/s44289-025-00060-x.

Evangelou, V. P. (Bill), & Zhang, Y. L. (1995). A review: Pyrite oxidation mechanisms and acid mine drainage prevention. *Critical Reviews in Environmental Science and Technology, 25*(2), 141–199. https://doi.org/10.1080/10643389509388477

Ezekwe, I. C, Ezekwe, A. S, & Chima, G. N (2013). Metal loadings and alkaline mine drainage from active and abandoned mines in the Ivo River Basin Area of Southeastern Nigeria. *Mine Water and the Environment, 32*(2), 97–107. doi:10.1007/s10230-013-0220-1.

Figuerola, B., Capdevila, P., Cerdà-Domènech, M., Garrabou, J., Mirasole, A., Bassols, P., del Campo, J., & Teixidó, N. ú (2025). Interactive effects of ocean acidification and warming disrupt calcification and microbiome composition in bryozoans. *Communications Biology, 8*(1). doi:10.1038/s42003-025-08524-8.

Fortin, J. L., Desmettre, T., Manzon, C., Judic-Peureux, V., Peugeot-Mortier, C., Giocanti, J. P., Hachelaf, M., Grangeon, M., Hostalek, U., Crouzet, J., & Capellier, G. (2010). Cyanide poisoning and cardiac disorders: 161 cases. *Journal of Emergency Medicine, 38*(4), 467–476. doi:10.1016/j.jemermed.2009.09.028.

Fowler, T. A., Holmes, P. R., & Crundwell, F. K. (2001). On the kinetics and mechanism of the dissolution of pyrite in the presence of Thiobacillus ferrooxidans. *Hydrometallurgy, 59*(2-3), 257–270. doi:10.1016/s0304-386x(00)00172-9.

Geng, M., & Duan, Z. (2010). Prediction of oxygen solubility in pure water and brines up to high temperatures and pressures. *Geochimica et Cosmochimica Acta, 74*(19), 5631–5640. https://doi.org/10.1016/j.gca.2010.06.034.

Gerke, H. H., Molson, J. W., & Frind, E. O. (1998). Modelling the effect of chemical heterogeneity on acidification and solute leaching in overburden mine spoils. *Journal of Hydrology, 209*(1–4), 166–185. https://doi.org/10.1016/S0022-1694(98)00106-1.

Geysels, B., Hiemstra, T., Groenenberg, J. E., & Comans, R. N. J. (2025). Glyphosate binding and speciation at the water-goethite interface: A surface complexation model consistent with IR spectroscopy and MO/DFT. *Water Research, 273*. Available from https://www.sciencedirect.com/science/journal/00431354 . doi:10.1016/j.watres.2024.123031.

Giantin, S., Franzin, A., Brusa, F., Montemurro, V., Bozzetta, E., Caprai, E., Fedrizzi, G., Girolami, F., & Nebbia, C. (2024). Overview of cyanide poisoning in cattle from Sorghum halepense and S. bicolor cultivars in Northwest Italy. *Animals, 14*(5), 743. doi:10.3390/ani14050743.

Gnann, S., Baldwin, J. W., Cuthbert, M. O., Gleeson, T., Schwanghart, W., & Wagener, T. (2025). The Influence of Topography on the Global Terrestrial Water Cycle. *Reviews of Geophysics, 63*(1), 1–40. https://doi.org/10.1029/2023RG000810.

Goddéris, Y., Schott, J., & Brantley, S. L. (2019). Reactive Transport Models of Weathering. *Elements, 15*(2), 103–106. https://doi.org/10.2138/gselements.15.2.103.

Goumih, A., El Adnani, M., Hakkou, R., Benzaazoua, M., Ouhamdouch, S., & Boumehdi, M. A. (2022). Evaluation of the long-term contaminated neutral drainage CND generation potential of waste rock piles at the abandoned Zn-Pb Erdouz Mine (Occidental High Atlas, Morocco). *Mining, Metallurgy and Exploration, 39*(2), 643–654. Available from https://link.springer.com/journal/volumesAndIssues/42461 . doi:10.1007/s42461-022-00553-9.

Goyne, K. W., Brantley, S. L., & Chorover, J. (2006). Effects of organic acids and dissolved oxygen on apatite and chalcopyrite dissolution: Implications for using elements as organomarkers and oxymarkers. *Chemical Geology, 234*(1-2), 28–45. doi:10.1016/j.chemgeo.2006.04.003.

Grimmett, Z. W., Schindler, J. C., & Stamler, J. S. (2025). Gases define redox signalling: NO, H2S, O2 … and cyanide: Redox signalling. *Nature Metabolism*, 7(3), 444–446. Available from https://www.nature.com/natmetab/ . doi:10.1038/s42255-025-01229-6.

Gunsinger, M. R., Ptacek, C. J., Blowes, D. W., Jambor, J. L., & Moncur, M. C. (2006). Mechanisms controlling acid neutralization and metal mobility within a Ni-rich tailings impoundment. *Applied Geochemistry, 21*(8), 1301–1321. doi:10.1016/j.apgeochem.2006.06.006.

Güven, O., Molz, F. J., & Melville, J. G. (1986). Comment on "An Advection-Diffusion Concept for Solute Transport in Heterogeneous Unconsolidated Geological Deposits. *Water Resources Research, 22*(1), 89–91. https://doi.org/10.1029/WR022i001p00089.

Gysi, A. P., & Williams-Jones, A. E. (2015). The thermodynamic properties of bastnäsite-(Ce) and parisite-(Ce). *Chemical Geology, 392*, 87–101. Available from http://www.sciencedirect.com/science/journal/00092541 . doi:10.1016/j.chemgeo.2014.11.001.

Hakkou, R., Benzaazoua, M., & Bussière, B. (2008). Acid mine drainage at the Abandoned Kettara Mine (Morocco): 1. Environmental characterization. *Mine Water and the Environment, 27*(3), 145–159. doi:10.1007/s10230-008-0036-6.

Hall, J. O. (2018). Nitrate- and Nitrite-Accumulating Plants. *Veterinary Toxicology: Basic and Clinical Principles: Third Edition (Third Edit)*. Elsevier Inc https://doi.org/10.1016/B978-0-12-811410-0.00065-9.

Haneklaus, N. H., Mwalongo, D. A., Lisuma, J. B., Amasi, A. I., Mwimanzi, J., Bituh, T., Ćirić, J., Nowak, J., Ryszko, U., Rusek, P., Maged, A., Bilal, E., Bellefqih, H., Qamouche, K., Brahim, J. A., Beniazza, R., Mazouz, H., van der Merwe, E. M., Truter, W., … Wacławek, S. (2024). Rare earth elements and uranium in Minjingu phosphate fertilizer products: Plant food for thought. *Resources, Conservation and Recycling, 207*. Available from https://www.sciencedirect.com/science/journal/09213449 . doi:10.1016/j.resconrec.2024.107694.

Hashimoto, S., Gojo, K., Hikota, S., Sendai, N., & Otsuki, A. (1999). Nitrous oxide emissions from coastal waters in Tokyo Bay. *Marine Environmental Research, 47*(3), 213–223. doi:10.1016/s0141-1136(98)00118-4.

Hassoun, A. E. R., Mojtahid, M., Merheb, M., Lionello, P., Gattuso, J. P., & Cramer, W. (2025). Climate change risks on key open marine and coastal mediterranean ecosystems. *Scientific Reports, 15*(1). Available from https://www.nature.com/srep/ . doi:10.1038/s41598-025-07858-x.

Heikkinen, P. M., & Räisänen, M. L. (2008). Mineralogical and geochemical alteration of Hitura sulphide mine tailings with emphasis on nickel mobility and retention. *Journal of Geochemical Exploration, 97*(1), 1–20. doi:10.1016/j.gexplo.2007.09.001.

Heikkinen, P. M., & Räisänen, M. L. (2009). Trace metal and As solid-phase speciation in sulphide mine tailings—Indicators of spatial distribution of sulphide oxidation in active tailings impoundments. *Applied Geochemistry, 24*(7), 1224–1237. doi:10.1016/j.apgeochem.2009.03.007.

Heikkinen, P. M., Räisänen, M. L., & Johnson, R. H. (2009). Geochemical characterisation of seepage and drainage water quality from two sulphide mine tailings impoundments: Acid mine drainage versus neutral mine drainage. *Mine Water and the Environment, 28*(1), 30–49. doi:10.1007/s10230-008-0056-2.

Hendry, M. J., Wassenaar, L. I., Barbour, S. L., Schabert, M. S., Birkham, T. K., Fedec, T., & Schmeling, E. E. (2018). Assessing the fate of explosives derived nitrate in mine waste rock dumps using the stable isotopes of oxygen and nitrogen. *Science of the Total Environment, 640-641*, 127–137. Available from www.elsevier.com/locate/scitotenv . doi:10.1016/j.scitotenv.2018.05.275.

Holden, F. J., Davies, K., Bird, M. I., Hume, R., Green, H., Beerling, D. J., & Nelson, P. N. (2024). In-field carbon dioxide removal via weathering of crushed basalt applied to acidic tropical agricultural soil. *Science of the Total Environment, 955*(September), 176568. https://doi.org/10.1016/j.scitotenv.2024.176568

Holmes, P. R., & Crundwell, F. K. (2000). The kinetics of the oxidation of pyrite by ferric ions and dissolved oxygen: An electrochemical study. *Geochimica et Cosmochimica Acta, 64*(2), 263–274. doi:10.1016/S0016-7037(99)00296-3.

Hossain, M., & Kumar Patra, P. (2020). Hydrogeochemical characterisation and health hazards of fluoride enriched groundwater in diverse aquifer types. *Environmental Pollution, 258*, 113646. doi:10.1016/j.envpol.2019.113646.

House, J. E., & House, K. A. (2016). Nitrogen. Descriptive Inorganic Chemistry (pp. 197–214). Elsevier. https://doi.org/10.1016/B978-0-12-804697-5.00013-0.

Hsiao, S. C., Wu, H. L., Fu, H. S., & Chen, W. B. (2025). Accelerated Ocean acidification (1985–2022) threatens tropical coral reefs and highlights biogeochemical refugia for marine conservation. *Journal of Sea Research, 207*. Available from https://www.sciencedirect.com/science/journal/13851101 . doi:10.1016/j.seares.2025.102612.

Hulver, A. M., Teixidó, N., Kemp, D. W., Keister, E. F., Gattuso, J. P., & Grottoli, A. G. (2025). High heterotrophic capacity favors Mediterranean coral success and resilience in the face of ocean acidification. *Coral Reefs, 44*(4), 1041–1054. Available from https://www.springer.com/journal/338 . doi:10.1007/s00338-025-02663-4.

Hunt, R. J., Walker, J. E., & Karabbenhoft, D. P. (1999). Characterizing hydrology and the importance of ground-water discharge in natural and constructed wetlands. *Wetlands, 19*(2), 458–472.

Idini, A., & Frau, F. (2025). The precursor of apatite: Octacalcium phosphate (OCP) in the earth and environmental sciences—A review. *Earth-Science Reviews, 261*, 105044. doi:10.1016/j.earscirev.2025.105044.

Ighalo, J. O., Kurniawan, S. B., Iwuozor, K. O., Aniagor, C. O., Ajala, O. J., Oba, S. N., Iwuchukwu, F. U., Ahmadi, S., & Igwegbe, C. A. (2022). A review of treatment technologies for the mitigation of the toxic environmental effects of acid mine drainage (AMD). *Process Safety and Environmental Protection, 157*, 37–58. Available from http://www.elsevier.com/wps/find/journaldescription.cws_home/713889/description#description . doi:10.1016/j.psep.2021.11.008.

Iwanoff, A. (1998). Environmental impacts of deep opencast limestone mines in Laegerdorf, Northern Germany. *Mine Water and the Environment, 17*(1), 52–61. doi:10.1007/bf02687244.

Jang, M., & Kwon, H. (2011). Pilot-scale tests to optimize the treatment of net-alkaline mine drainage. *Environmental Geochemistry and Health, 33*(S1), 91–101. doi:10.1007/s10653-010-9353-3.

Jarvis, A. P., Gandy, C. J., & Webb, J. A. (2023). Controls on the generation and geochemistry of neutral mine drainage: Evidence from Force Crag Mine, Cumbria, UK. *Minerals, 13*(5). Available from http://www.mdpi.com/journal/minerals . doi:10.3390/min13050592.

Jasani, S. (2015). Smoke inhalation. *Small animal critical care medicine* (pp. 785–788). Elsevier. https://doi.org/10.1016/B978-1-4557-0306-7.00147-1.

Jaszczak, E., Polkowska, Ż., Narkowicz, S., & Namieśnik, J. (2017). Cyanides in the environment—analysis—problems and challenges. *Environmental Science and Pollution Research, 24*(19), 15929–15948. https://doi.org/10.1007/s11356-017-9081-7.

Jiao, Y., Zhang, C., Su, P., Tang, Y., Huang, Z., & Ma, T. (2023). A review of acid mine drainage: Formation mechanism, treatment technology, typical engineering cases and resource utilization. *Process Safety and Environmental Protection, 170*, 1240–1260. doi:10.1016/j.psep.2022.12.083.

Jing, W., Debang, Y., Qiao, H., Lu, L., Hang, J., Yinfei, Q., Kang, N., & Yi, C. (2025). Alleviation of soil acidification using livestock manure increases nitrogen supply while stimulating N2O emission in a tea plantation. *Archives of Agronomy and Soil Science, 71*(1), 1–13. Available from http://www.tandf.co.uk/journals/titles/03650340.asp . doi:10.1080/03650340.2025.2555673.

Jones, K., Liess, A., & Sjöstedt, J. (2025). Microbial carbon utilization in a boreal lake under the combined pressures of brownification and eutrophication: Insights from a field experiment. *Hydrobiologia, 852*(3), 721–733. doi:10.1007/s10750-024-05718-9.

Jørgensen, B. B. (1982). Mineralization of organic matter in the sea bed - The role of sulphate reduction. *Nature, (Vol. 296, Issue 5858*, 643–645. https://doi.org/10.1038/296643a0.

Jung, Y.-Y., Choi, S.-H., Choi, M., Bong, Y.-S., Park, M.-Y., Lee, K.-S., & Shin, W.-J. (2023). Acid mine drainage and smelter-derived sources affecting water geochemistry in the upper Nakdong River, South Korea. *Science of The Total Environment, 880*, 163353. doi:10.1016/j.scitotenv.2023.163353.

Kaaya, N. I., Vegi, M. R., & Macheyeki, A. S. (2025). Graphene-based adsorbents for selective recovery of rare earth elements from mining wastes: A review. *FlatChem, 54*. Available from https://www.sciencedirect.com/science/journal/24522627 . doi:10.1016/j.flatc.2025.100954.

Kaaya, N. I., Vegi, M. R., & Macheyeki, A. S. (2026). Acid mine drainage, soil pollution, and carcinogenic risk of geogenic contaminants in artisanal and small-scale gold mining areas of Geita, Tanzania. *Environmental Geochemistry and Health, 48*(1), 23. https://doi.org/10.1007/s10653-025-02909-8.

Kalonji Kabambi, A., Bussière, B., & Demers, I. (2017). Hydrogeological Behaviour of Covers with Capillary Barrier Effects Made of Mining Materials. *Geotechnical and Geological Engineering, 35*(3), 1199–1220. https://doi.org/10.1007/s10706-017-0174-3.

Kang, C., Pan, J., Luo, Y., Xie, X., Yang, Y., Zou, Z., Zhao, X., Chen, M., & Wang, C. (2025). Dual-functional remediation of ammonia nitrogen contamination in ionic rare earth mines: Efficient washing and soil stabilization using lignosulfonates. *Environmental Technology and Innovation, 39*(March), 104297. https://doi.org/10.1016/j.eti.2025.104297.

Kang, C.-U., Jeon, B.-H., Park, S.-S., Kang, J.-S., Kim, K.-H., Kim, D.-K., Choi, U.-K., & Kim, S.-J. (2015). Inhibition of pyrite oxidation by surface coating: A long-term field study. *Environmental Geochemistry and Health*. https://doi.org/10.1007/s10653-015-9778-9.

Kant, C., Meena, R. S., & Singh, S. K. (2025). A Critical Appraisal on Various Hydrological and Hydrodynamic Models. *Water Conservation Science and Engineering, 10*(1). https://doi.org/10.1007/s41101-024-00328-x.

Keays, R. R., & Lightfoot, P. C. (2010). Crustal sulfur is required to form magmatic Ni-Cu sulfide deposits: Evidence from chalcophile element signatures of Siberian and Deccan Trap basalts. *Mineralium Deposita, 45*(3), 241–257. doi:10.1007/s00126-009-0271-1.

Kidavu, A. VS., Chaudhary, A. K, & Soni, P. K (2025). Study of thermal decomposition routes and detection of explosives from defense test range soil samples using a tunable mid IR QCL laser based photoacoustic thermal analyzer. *Measurement, 248*, 116791. doi:10.1016/j.measurement.2025.116791.

Kim, J. O., Lee, Y. W., & Chung, J. (2013). The role of organic acids in the mobilization of heavy metals from soil. *KSCE Journal of Civil Engineering, 17*(7), 1596–1602. doi:10.1007/s12205-013-0323-z.

Kohio, E. N., Sossou, S. K., Karoui, H., & Yacouba, H. (2025). Environmental Cyanide Pollution from Artisanal Gold Mining in Burkina Faso: Human Exposure Risk Analysis Based on a Conceptual Site Model. *International Journal of Environmental Research and Public Health, 22*(7). https://doi.org/10.3390/ijerph22071125.

Kumar, A., Shemi, A., Chipise, L., Yah, C. S., Moodley, S., & Ndlovu, S. (2025). Optimizing biogenic cyanide production using indigenous cyanogenic microorganisms for bioleaching of precious metals. *Journal of Sustainable Metallurgy, 11*(4), 3700–3715. Available from https://www.springer.com/journal/40831 . doi:10.1007/s40831-025-01110-6.

Kusi, J. K., Foli, G., Peasah, M. Y., & Akoto, O. (2024). Geochemical characterization of waste rock dump and its potential to predict acid mine drainage. *Scientific African, 26*, e02350. doi:10.1016/j.sciaf.2024.e02350.

Kuyucak, N., & Akcil, A. (2013). Cyanide and removal options from effluents in gold mining and metallurgical processes. *Minerals Engineering, 50-51*, 13–29. doi:10.1016/j.mineng.2013.05.027.

Leharne, S. (2019). Transfer phenomena and interactions of non-aqueous phase liquids in soil and groundwater. *ChemTexts, 5*(1), 5. https://doi.org/10.1007/s40828-019-0079-2.

Lehmann, K., Arachchige, D. E., Lehmann, R., Overholt, W. A., Küsel, K., & Totsche, K. U. (2026). Neglected but pivotal: Complex matter dynamics in the aeration zone contribute to groundwater quality evolution. *Water Research, 291*(December 2025), 125287. https://doi.org/10.1016/j.watres.2025.125287.

Lehner, S., Savage, K., Ciobanu, M., & Cliffel, D. E. (2007). The effect of As, Co, and Ni impurities on pyrite oxidation kinetics: An electrochemical study of synthetic pyrite. *Geochimica et Cosmochimica Acta, 71*(10), 2491–2509. doi:10.1016/j.gca.2007.03.005.

Lindsay, M. B. J., Condon, P. D., Jambor, J. L., Lear, K. G., Blowes, D. W., & Ptacek, C. J. (2009). Mineralogical, geochemical, and microbial investigation of a sulfide-rich tailings deposit characterized by neutral drainage. *Applied Geochemistry, 24*(12), 2212–2221. doi:10.1016/j.apgeochem.2009.09.012.

Lindsay, M. B. J., Moncur, M. C., Bain, J. G., Jambor, J. L., Ptacek, C. J., & Blowes, D. W. (2015). Geochemical and mineralogical aspects of sulfide mine tailings. *Applied Geochemistry, 57*, 157–177. Available from http://www.journals.elsevier.com/applied-geochemistry . doi:10.1016/j.apgeochem.2015.01.009.

Liu, F., Qiao, X., Zhou, L., & Zhang, J. (2018). Migration and fate of acid mine drainage pollutants in calcareous soil, *International Journal of Environmental Research and Public Health* p. 1759. https://doi.org/10.3390/ijerph15081759.

Liu, R., Liu, S., Bai, X., Liu, S., & Liu, Y. (2025). Biooxidation of Arsenopyrite by Acidithiobacillus ferriphilus QBS 3 exhibits arsenic resistance under extremely acidic bioleaching conditions. *Biology, 14*(5), 550. doi:10.3390/biology14050550.

Liu, S. L., Fan, H. R., Liu, X., Meng, J., Butcher, A. R., Yann, L., Yang, K. F., & Li, X. C. (2023). Global rare earth elements projects: New developments and supply chains. *Ore Geology Reviews, 157*. Available from http://www.sciencedirect.com/science/journal/01691368 . doi:10.1016/j.oregeorev.2023.105428.

Liu, X., Zhao, K., Zhang, Y., Xu, X., & Miao, X. (2025). Lateral photovoltage in natural pyrite: Mechanism, model, and characteristic. *Materials Today Electronics, 14*, 100183. https://doi.org/10.1016/j.mtelec.2025.100183.

Liu, Y., Yang, W., Wu, Y., Cai, W. J., Li, C., Lin, H., Zhuang, P., Zhang, J., Xu, Y., Qiu, H., Huang, Y., & Qi, D. (2025). Emergent seasonal hypoxia and acidification risks induced by seaweed and fish polyculture in the world's largest seaweed farm. *Limnology and Oceanography Letters, 10*(6), 911–922. Available from https://aslopubs.onlinelibrary.wiley.com/loi/23782242 . doi:10.1002/lol2.70055.

Madinabeitia, S. G. D. e, Velasco-acebes, J., Ibarguchi, J. I. G., & Layne, G. (2025). Fe, Cu and S isotopes as tracers of microbial reduction in the shale-hosted VMS deposits of the Iberian Pyrite Belt: 277 October 2024 https://doi.org/10.1016/j.gexplo.2025.107819 .

Mafra, C., Bouzahzah, H., Stamenov, L., & Gaydardzhiev, S. (2020). Insights on the effect of pyrite liberation degree upon the acid mine drainage potential of sulfide flotation tailings. *Applied Geochemistry, 123*(March), 104774. https://doi.org/10.1016/j.apgeochem.2020.104774.

Mansur, E. T., Barnes, S. J., & Duran, C. J. (2021). An overview of chalcophile element contents of pyrrhotite, pentlandite, chalcopyrite, and pyrite from magmatic Ni-Cu-PGE sulfide deposits. *Mineralium Deposita, 56*(1), 179–204. Available from http://link.springer.de/link/service/journals/00126/index.htm . doi:10.1007/s00126-020-01014-3.

Marshall, B. G., Veiga, M. M., da Silva, H. A. M., & Guimarães, J. R. D. (2020). Cyanide contamination of the Puyango-Tumbes River caused by Artisanal Gold Mining in Portovelo-Zaruma, Ecuador. *Current Environmental Health Reports, 7*(3), 303–310. Available from http://link.springer.com/journal/40572 . doi:10.1007/s40572-020-00276-3.

Masindi, V., Chatzisymeon, E., Kortidis, I., & Foteinis, S. (2018). Assessing the sustainability of acid mine drainage (AMD) treatment in South Africa. *Science of the Total Environment, 635*, 793–802. https://doi.org/10.1016/j.scitotenv.2018.04.108.

Masindi, V., Chatzisymeon, E., Kortidis, I., & Foteinis, S. (2018). Assessing the sustainability of acid mine drainage (AMD) treatment in South Africa. *Science of the Total Environment, 635*, 793–802. doi:10.1016/j.scitotenv.2018.04.108.

Masindi, V., Foteinis, S., Renforth, P., Ndiritu, J., Maree, J. P., Tekere, M., & Chatzisymeon, E. (2022). Challenges and avenues for acid mine drainage treatment, beneficiation, and valorisation in circular economy: A review. *Ecological Engineering, 183*, 106740. doi:10.1016/j.ecoleng.2022.106740.

Mather, T. A., Allen, A. G., Davison, B. M., Pyle, D. M., Oppenheimer, C., & McGonigle, A. J. S. (2004). Nitric acid from volcanoes. *Earth and Planetary Science Letters, 218*(1-2), 17–30. doi:10.1016/s0012-821x(03)00640-x.

Mehdaoui, H. Y., Jouini, M., Lefebvre, J., Neculita, C. M., Pabst, T., & Benzaazoua, M. (2025). Geochemical stability of As-rich residues from a 20-year-old passive field biofilter used for neutral mine drainage treatment. *Ecological Engineering, 216*. Available from https://www.sciencedirect.com/science/journal/09258574 . doi:10.1016/j.ecoleng.2025.107618.

Meutia, A. A., Bachriadi, D., & Gafur, N. A. (2023). Environment Degradation, Health Threats, and Legality at the Artisanal Small-Scale Gold Mining Sites in Indonesia. *International Journal of Environmental Research and Public Health, 20*(18). https://doi.org/10.3390/ijerph20186774.

Mhangara, P., Matyukira, C., & Otunola, B. O. (2026). Digging deep: the transformation of penhalonga's landscape through artisanal mining. *Environmental Earth Sciences, 85*(1), 1–13. https://doi.org/10.1007/s12665-025-12668-5.

Michael, A., & Varnava, N. (2024). Ioannis Ioannidis, Ioannis Pashalidis, Uranium recovery from phosphate rocks/minerals–A comprehensive review. *Sustainable Chemistry for the Environment, 5*, 100055. doi:10.1016/j.scenv.2023.100055.

Miller, J. D., Li, J., Davidtz, J. C., & Vos, F. (2005). A review of pyrrhotite flotation chemistry in the processing of PGM ores. *Minerals Engineering, 18*(8), 855–865. doi:10.1016/j.mineng.2005.02.011.

Modabberi, S., Alizadegan, A., & Mirnejad, H. (2013). Esmat Esmaeilzadeh, Prediction of AMD generation potential in mining waste piles, in the sarcheshmeh porphyry copper deposit, Iran. *Environmental Monitoring and Assessment, 185*(11), 9077–9087. doi:10.1007/s10661-013-3237-9.

Moncur, M. C., Jambor, J. L., Ptacek, C. J., & Blowes, D. W. (2009). Mine drainage from the weathering of sulfide minerals and magnetite. *Applied Geochemistry, 24*(12), 2362–2373. doi:10.1016/j.apgeochem.2009.09.013.

Moore, J., Lichtner, P. C., White, A. F., & Brantley, S. L. (2012). Using a reactive transport model to elucidate differences between laboratory and field dissolution rates in regolith. *Geochimica et Cosmochimica Acta, 93*, 235–261. https://doi.org/10.1016/j.gca.2012.03.021.

Musaogullari, A., & Chai, Y.-C. (2020). Redox regulation by protein S-glutathionylation: From molecular mechanisms to implications in health and disease. *International Journal of Molecular Sciences, 21*(21), 8113. doi:10.3390/ijms21218113.

Naldrett, A. J. (1992). A model for the Ni-Cu-PGE ores of the Noril'sk region and its application to other areas of flood basalt. *Economic Geology, 87*(8), 1945–1962. doi:10.2113/gsecongeo.87.8.1945.

Naldrett, A. J. (2013). *Magmatic sulfide deposits: Geology, geochemistry and exploration*. Springer Science & Business Media.

Naveed, M., Abid, I., Munir, M., Ghafoor, A., Alqahtani, N., Islam, S., Ali-Dinar, H., & Mustafa, A. (2025). Precision-based augmented manure for pH-targeted soil acidification: Improved phosphate-mediated cadmium immobilization and canola growth in calcareous soils. *Chemical and Biological Technologies in Agriculture, 12*(1). doi:10.1186/s40538-025-00883-y.

Nayak, N. P., Kudapa, V. K., Bhan, U., Goswami, L., Kumar, S., & Kushwaha, A. (2023). Impacts of Climate Change on Weathering and Erosion of Rock Types Exposed on Earths Surface. *Weathering and Erosion Processes in the Natural Environment* (pp. 359–373). Wiley. https://doi.org/10.1002/9781394157365.ch15.

Nordstrom, D.K. (2013). Extremely arsenic-rich, pH-neutral waters from the Giant Mine, Canada. https://pubs.usgs.gov/publication/70190120.

Nordstrom, D. K., Alpers, C. N., Ptacek, C. J., & Blowes, D. W. (2000). Negative pH and extremely acidic mine waters from Iron Mountain, California. *Environmental Science and Technology, 34*(2), 254–258. doi:10.1021/es990646v.

Nordstrom, D. K., Blowes, D. W., & Ptacek, C. J. (2015). Hydrogeochemistry and microbiology of mine drainage: An update. *Applied Geochemistry, 57*, 3–16. Available from http://www.journals.elsevier.com/applied-geochemistry . doi:10.1016/j.apgeochem.2015.02.008.

Norris, P. R., Burton, N. P., & Clark, D. A. (2013). Mineral sulfide concentrate leaching in high temperature bioreactors. *Minerals Engineering, 48*, 10–19. https://doi.org/10.1016/j.mineng.2013.01.001.

Norton, J., & Ouyang, Y. (2019). Controls and adaptive management of nitrification in agricultural soils. *Frontiers in Microbiology, 10*. doi:10.3389/fmicb.2019.01931.

Olds, W. E., Weber, P. A., Pizey, M. H., & Pope, J. (2016). Acid mine drainage analysis for the Reddale Coal Mine, Reefton, New Zealand. *New Zealand Journal of Geology and Geophysics, 59*(2), 341–351. Available from http://www.tandfonline.com/loi/tnzg20?open=56&repitition=0#vol_56 . doi:10.1080/00288306.2016.1148056.

Oliveira, M., & Machado, A. V. (2013). The role of phosphorus on eutrophication: A historical review and future perspectives. *Environmental Technology Reviews, 2*(1), 117–127. doi:10.1080/21622515.2013.861877.

Olson, E. M., John, J. G., Dunne, J. P., Stock, C. A., & Drenkard, E. J. (2025). Potential for regional resilience to ocean warming and acidification extremes: Projected vulnerability under contrasting pathways and thresholds. *Global Change Biology, 31*(7). Available from http://onlinelibrary.wiley.com/journal/10.1111/ . doi:10.1111/gcb.70360. (ISSN)1365-2486.

Orndorff, Z. W., & Daniels, W. L. (2004). Evaluation of acid-producing sulfidic materials in Virginia highway corridors. *Environmental Geology, 46*(2), 209–216. Available from link.springer.de/link/service/journals/00254/index.htm . doi:10.1007/s00254-004-1027-y.

Oruc, H. H., Akkoc, A., Uzunoglu, I., & Kennerman, E. (2010). Nitrate Poisoning in Horses Associated With Ingestion of Forage and Alfalfa. *Journal of Equine Veterinary Science, 30*(3), 159–162. https://doi.org/10.1016/j.jevs.2010.01.055.

Osellame, L. D., Blacker, T. S., & Duchen, M. R. (2012). Cellular and molecular mechanisms of mitochondrial function. *Best Practice and Research: Clinical Endocrinology and Metabolism, 26*(6), 711–723. Available from http://www.elsevier.com/inca/publications/store/6/2/3/0/0/1/index.htt . doi:10.1016/j.beem.2012.05.003.

Oseyemi, A. E, Zabihihesari, A., Salahandish, R., & Rezai, P. (2025). Low-level salinity monitoring in drinking water using an electrochemical microfluidic sensor with an ion-selective polymer membrane. *Microchemical Journal, 215*, 114497. doi:10.1016/j.microc.2025.114497.

Parkhurst, D. L., & Appelo, C. A. J. (1999). User's guide to PHREEQC (ver2) - A computer program for speciation, batch-reaction, one-dimensional transport, and inverse geochemical calculations. *U.S. Geological Survey 1999, 99–4259*(xiv), 312. https://pubs.usgs.gov/wri/1999/4259/report.pdf

Pakostova, E., Hilger, D. M., Blowes, D. W., & Ptacek, C. J. (2023). Microbial processes with the potential to mobilize As from a circumneutral-pH mixture of flotation and roaster tailings. *Scientific Reports, 13*(1). Available from https://www.nature.com/srep/ . doi:10.1038/s41598-023-50435-3.

Paktunc, A. D. (1999). Discussion of "A method to calculate the neutralization potential of mining wastes" by Lawrence and Scheske. *Environmental Geology, 38*(1), 82–84. doi:10.1007/s002540050404.

Pantsar-Kallio, M., & Manninen, P. K. G. (1997). Speciation of mobile arsenic in soil samples as a function of pH. *Science of the Total Environment, 204*(2), 193–200. doi:10.1016/S0048-9697(97)00176-9.

Patra, S., & Punekar, J. (2025). Ocean acidification signals through deep time: A review of proxies. *Evolving Earth, 3*, 100056. doi:10.1016/j.eve.2024.100056.

Peppas, A., Komnitsas, K., & Halikia, I. (2000). Use of organic covers for acid mine drainage control. *Minerals Engineering, 13*(5), 563–574. https://doi.org/10.1016/S0892-6875(00)00036-4.

Petrikovics, I., Budai, M., Kovacs, K., & Thompson, D. E (2015). Past, present and future of cyanide antagonism research: From the early remedies to the current therapies. *World Journal of Methodology, 5*(2), 88. doi:10.5662/wjm.v5.i2.88.

Phillips, F. M., & Castro, M. C. (2013). Groundwater Dating and Residence-Time Measurements. Treatise on Geochemistry: 7 (Second Edition (2nd ed). Elsevier Ltd. https://doi.org/10.1016/B978-0-08-095975-7.00513-1.

Picard, A., Gartman, A., Clarke, D. R., & Girguis, P. R. (2018). Sulfate-reducing bacteria influence the nucleation and growth of mackinawite and greigite. *Geochimica et Cosmochimica Acta, 220*, 367–384. Available from http://www.journals.elsevier.com/geochimica-et-cosmochimica-acta/ . doi:10.1016/j.gca.2017.10.006.

Pjetraj, D., Mechri, M. El, Bacelli, S., Fabiani, E., Caponi, L., Gatti, S., & Lionetti, E (2025). A blue blood toddler– a case report of methemoglobinemia and literature review. *Italian Journal of Pediatrics, 51*(1), 53. https://doi.org/10.1186/s13052-025-01886-z.

Plante, B., Benzaazoua, M., & Bussière, B. (2011a). Kinetic testing and sorption studies by modified weathering cells to characterize the potential to generate contaminated neutral drainage. *Mine Water and the Environment, 30*(1), 22–37. https://doi.org/10.1007/s10230-010-0131-3.

Plante, B., Benzaazoua, M., & Bussière, B. (2011b). Predicting geochemical behaviour of waste rock with low acid generating potential using laboratory kinetic tests. *Mine Water and the Environment, 30*(1), 2–21. https://doi.org/10.1007/s10230-010-0127-z.

Plante, B., Benzaazoua, M., Bussière, B., Biesinger, M. C., & Pratt, A. R. (2010). Study of Ni sorption onto Tio mine waste rock surfaces. *Applied Geochemistry, 25*(12), 1830–1844. doi:10.1016/j.apgeochem.2010.09.010.

Plante, B., Bussière, B., & Benzaazoua, M. (2014). Lab to field scale effects on contaminated neutral drainage prediction from the Tio mine waste rocks. *Journal of Geochemical Exploration, 137*, 37–47. https://doi.org/10.1016/j.gexplo.2013.11.004.

Qian, Y., Zhao, Y., Wu, Q. l., & Yang, Y. (2018). Review of salinity measurement technology based on optical fiber sensor. *Sensors and Actuators, B: Chemical, 260*, 86–105. doi:10.1016/j.snb.2017.12.077.

Qiu, J., Tao, H., Li, H., Liu, X., Liu, R., Nawaz, M. N., Wang, X., & Ma, L. (2025). The environmental adaptation of acidophilic archaea: Promotion of horizontal gene transfer by genomic islands. *BMC Genomics, 26*(1). Available from https://bmcgenomics.biomedcentral.com/ . doi:10.1186/s12864-025-11875-5.

Queizán, M., Graña, A. M., & Hermida-Ramón, J. M. (2022). Achieving S2- in aqueous Solution: An evaluation using First-Principle molecular dynamics simulations. *Journal of Molecular Liquids, 349*, 2–8. https://doi.org/10.1016/j.molliq.2021.118109.

Raiswell, R., & Berner, R. A. (1986). Pyrite and organic matter in Phanerozoic normal marine shales. *Geochimica et Cosmochimica Acta, 50*(9), 1967–1976. doi:10.1016/0016-7037(86)90252-8.

Rao, A. S., Jha, P., Meena, B. P., Biswas, A. K., Lakaria, B. L., & Patra, A. K. (2017). *Nitrogen processes in agroecosystems of India The Indian nitrogen assessment: Sources of reactive nitrogen, environmental and climate effects, management options, and policies* (pp. 59–76). India: Elsevier Available from. http://www.sciencedirect.com/science/book/9780128118368 . doi:10.1016/B978-0-12-811836-8.00005-7.

Radková, B. A., Jamieson, H. E., & Campbell, K. M. (2020). Antimony mobility during the early stages of stibnite weathering in tailings at the Beaver Brook Sb deposit, Newfoundland. *Applied Geochemistry, 115*(August 2019), 104528. https://doi.org/10.1016/j.apgeochem.2020.104528

Rice, K. C., & Herman, J. S. (2012). Acidification of earth: An assessment across mechanisms and scales. *Applied Geochemistry, 27*(1), 1–14. doi:10.1016/j.apgeochem.2011.09.001.

Rickard, D. (2019). How long does it take a pyrite framboid to form? *Earth and Planetary Science Letters, 513*, 64–68. doi:10.1016/j.epsl.2019.02.019.

Rimstidt, D. D., & Vaughan, D. J. (2003). Pyrite oxidation: A state-of-the-art assessment of the reaction mechanism. *Geochimica et Cosmochimica Acta, 67*(5), 873–880. doi:10.1016/S0016-7037(02)01165-1.

Rohwerder, T., Schippers, A., & Sand, W. (1998). Determination of reaction energy values for biological pyrite oxidation by calorimetry. *Thermochimica Acta, 309*(1–2), 79–85. https://doi.org/10.1016/s0040-6031(97)00352-3.

Romero-Mujalli, G., Hartmann, J., Börker, J., Gaillardet, J., & Calmels, D. (2019). Ecosystem controlled soil-rock pCO2 and carbonate weathering – Constraints by temperature and soil water content. *Chemical Geology, 527(October, 2017*, 118634. https://doi.org/10.1016/j.chemgeo.2018.01.030.

Rulis, P., Ouyang, L., & Ching, W. Y. (2004). Electronic structure and bonding in calcium apatite crystals: Hydroxyapatite, fluorapatite, chlorapatite, and bromapatite. *Physical Review B—Condensed Matter and Materials Physics, 70*(15), 1–155104. doi:10.1103/PhysRevB.70.155104.

Saadaoui, J., Smyth, R., & Vespignani, J. (2025). Ensuring the security of the clean energy transition: Examining the impact of geopolitical risk on the price of critical minerals. *Energy Economics, 142*, 108195. doi:10.1016/j.eneco.2025.108195.

Sarmiento, A. M., Grande, J. A., Luís, A. T., Dávila, J. M., Fortes, J. C., Santisteban, M., Curiel, J., de la Torre, M. L., & da Silva, E. F. (2018). Negative pH values in an open-air radical environment affected by acid mine drainage. Characterization and proposal of a hydrogeochemical model. *Science of the Total Environment, 644*, 1244–1253. Available from www.elsevier.com/locate/scitotenv . doi:10.1016/j.scitotenv.2018.06.381.

Sarmiento, A. M., Rodriguez-perez, A. M., Davila, J. M., Luisa, M., & Torre, D. (2024). Effect of acid mine drainage on the mechanical properties of AISI 1020 carbon steel and AW6060 aluminium. *Journal of Materials Research and Technology, 32*(July), 541–549. https://doi.org/10.1016/j.jmrt.2024.07.099.

Savage, K. S., Stefan, D., & Lehner, S. W. (2008). Impurities and heterogeneity in pyrite: Influences on electrical properties and oxidation products. *Applied Geochemistry, 23*(2), 103–120. doi:10.1016/j.apgeochem.2007.10.010.

Schippers, A. (2004). Biogeochemistry of metal sulfide oxidation in mining environments, sediments, and soils. In *Sulfur Biogeochemistry* -. Past and Present ((Vol. 379,, pp. 49–62). Geological Society of America. https://doi.org/10.1130/0-8137-2379-5.49.

Schoonen, M. A. A. (2004). Mechanisms of sedimentary pyrite formation. In *Sulfur Biogeochemistry* -. Past and Present ((Vol. 379,, pp. 117–134). Geological Society of America. https://doi.org/10.1130/0-8137-2379-5.117.

Schwab, A. P., Zhu, D. S., & Banks, M. K. (2008). Influence of organic acids on the transport of heavy metals in soil. *Chemosphere, 72*(6), 986–994. doi:10.1016/j.chemosphere.2008.02.047.

Shaji, E., Sarath, K. V., Santosh, M., Krishnaprasad, P. K., Arya, B. K., & Babu, M. S. (2024). Fluoride contamination in groundwater: A global review of the status, processes, challenges, and remedial measures. *Geoscience Frontiers, 15*(2), 101734. doi:10.1016/j.gsf.2023.101734.

Shangguan, Y., Li, B., Zhuang, X., Querol, X., Moreno, N., Huang, P., Guo, Y., Shi, Y., Wu, T., & Sola, P. C. (2025). Arsenic distribution and speciation in deposited coal mine dust. *Journal of Hazardous Materials, 482*. Available from https://www.sciencedirect.com/science/journal/03043894 . doi:10.1016/j.jhazmat.2024.136537.

Simon, A. C., & Ripley, E. M. (2011). The role of magmatic sulfur in the formation of ore deposits. *Reviews in Mineralogy and Geochemistry, 73*(1), 513–578. doi:10.2138/rmg.2011.73.16.

Singh, S., Anil, A. G., Kumar, V., Kapoor, D., Subramanian, S., Singh, J., & Ramamurthy, P. C. (2022). Nitrates in the environment: A critical review of their distribution, sensing techniques, ecological effects and remediation. *Chemosphere, 287*. Available from www.elsevier.com/locate/chemosphere . doi:10.1016/j.chemosphere.2021.131996.

Skousen, J., Renton, J., Brown, H., Evans, P., Leavitt, B., Brady, K., Cohen, L., & Ziemkiewicz, P. (1997). Neutralization potential of overburden samples containing siderite. *Journal of Environmental Quality, 26*(3), 673–681. doi:10.2134/jeq1997.00472425002600030012x.

Skousen, J., Zipper, C. E., Rose, A., Ziemkiewicz, P. F., Nairn, R., McDonald, L. M., & Kleinmann, R. L. (2017). Review of passive systems for acid mine drainage treatment. *Mine Water and the Environment, 36*(1), 133–153. Available from http://link.springer.de/link/service/journals/10230/index.htm . doi:10.1007/s10230-016-0417-1.

Skousen, J. G., Ziemkiewicz, P. F., & McDonald, L. M. (2019). Acid mine drainage formation, control and treatment: Approaches and strategies. *Extractive Industries and Society, 6*(1), 241–249. Available from http://www.journals.elsevier.com/the-extractive-industries-and-society/ . doi:10.1016/j.exis.2018.09.008.

Slater, L., & Comas, X. (2008). The Contribution of Ground Penetrating Radar to Water Resource Research. Ground Penetrating Radar Theory and Applications (First Edit). Elsevier https://doi.org/10.1016/B978-0-444-53348-7.00007-7.

Söderberg, T. U., Kleja, D. B., Åström, M., Jarsjö, J., Fröberg, M., Svensson, A., & Augustsson, A. (2019). Metal solubility and transport at a contaminated landfill site – From the source zone into the groundwater. *Science of The Total Environment, 668*, 1064–1076. https://doi.org/10.1016/j.scitotenv.2019.03.013.

Song, Q., & Yanful, E. K. (2011). Oxygen influx and geochemistry of percolate water from reactive mine waste rock underlying a sloping channelled soil cover. *Applied Geochemistry, 26*(5), 655–665. https://doi.org/10.1016/j.apgeochem.2011.01.024.

Sracek, O., Kříbek, B., Mihaljevič, M., Ettler, V. ě, Vaněk, A. š, Penížek, V. ít, Filip, J., Veselovský, F. š, & Nyambe, I. (2018). The impact of wetland on neutral mine drainage from mining wastes at Luanshya in the Zambian Copperbelt in the framework of climate change. *Environmental Science and Pollution Research, 25*(29), 28961–28972. doi:10.1007/s11356-018-2929-7.

Steadman, J. A., Large, R. R., Olin, P. H., Danyushevsky, L. V., Meffre, S., Huston, D., Fabris, A., Lisitsin, V., & Wells, T. (2021). Pyrite trace element behavior in magmatic-hydrothermal environments: An LA-ICPMS imaging study. *Ore Geology Reviews, 128*. Available from http://www.sciencedirect.com/science/journal/01691368 . doi:10.1016/j.oregeorev.2020.103878.

Stewart, J. A., Williams, B., LaVigne, M., Wanamaker, A. D., Strong, A. L., Jellison, B., Whitney, N. M., Thatcher, D. L., Robinson, L. F., Halfar, J., & Adey, W. (2025). Delayed onset of ocean acidification in the Gulf of Maine. *Scientific Reports, 15*(1). Available from https://www.nature.com/srep/ . doi:10.1038/s41598-024-84537-3.

Štyriaková, I., Štyriak, I., & Oberhänsli, H. (2012). *Rock weathering by indigenous heterotrophic bacteria of Bacillus spp. at different temperature: A laboratory experiment Mineralogy and Petrology, 105*, 135–144. https://doi.org/10.1007/s00710-012-0201-2.

Sui, C., Fatichi, S., Burlando, P., Weber, E., & Battista, G. (2022). Modeling distributed metal pollution transport in a mine impacted catchment: Short and long-term effects. *Science of the Total Environment, 812*, 151473. https://doi.org/10.1016/j.scitotenv.2021.151473.

Sweeney, R. E., & Kaplan, I. R. (1973). Pyrite framboid formation; laboratory synthesis and marine sediments. *Economic Geology, 68*(5), 618–634.

Tang, X., Chen, Y., Dai, Z., Chang, S. X., Zeng, R., He, D., & Xu, J. (2025). Multiple environmental stressors interactively affect soil phosphorus cycling microbiomes. *Communications Earth and Environment, 6*(1). Available from https://www.nature.com/commsenv/ . doi:10.1038/s43247-025-02772-6.

Third, K. A., Cord-Ruwisch, R., & Watling, H. R. (2000). The role of iron-oxidizing bacteria in stimulation or inhibition of chalcopyrite bioleaching. *Hydrometallurgy, 57*(3), 225–233. doi:10.1016/s0304-386x(00)00115-8.

Tu, Z., Wan, J., Guo, C., Fan, C., Zhang, T., Lu, G., Reinfelder, J. R., & Dang, Z. (2017). Electrochemical oxidation of pyrite in pH 2 electrolyte. *Electrochimica Acta, 239*, 25–35. Available from http://www.journals.elsevier.com/electrochimica-acta/ . doi:10.1016/j.electacta.2017.04.049.

Tulisiak, L. N., Loesing, M. E., Ljungberg, A., & Kaczor, E. E. (2025). Smartwatch Detection of Refractory Hypoxemia in Phenazopyridine-Induced Methemoglobinemia: A Case Report. *Journal of Medical Toxicology, 21*(4), 430–433. https://doi.org/10.1007/s13181-025-01090-9.

Vaughan, D. J., & Corkhill, C. L. (2017). Mineralogy of sulfides. *Elements, 13*(2), 81–87. Available from http://www.elementsmagazine.org/archivearticles/e13_2/e13_2.pdf . doi:10.2113/gselements.13.2.81.

Vriens, B., Plante, B., Seigneur, N., & Jamieson, H. (2020). Mine waste rock: Insights for sustainable hydrogeochemical management. *Minerals, 10*(9). doi:10.3390/min10090728.

Vriens, B., Skierszkan, E. K., St-Arnault, M., Salzsauler, K., Aranda, C., Mayer, K. U., & Beckie, R. D. (2019). Mobilization of Metal(oid) oxyanions through circumneutral mine waste-rock drainage. *ACS Omega, 4*(6), 10205–10215. Available from pubs.acs.org/journal/acsodf . doi:10.1021/acsomega.9b01270.

Vuhahula, E. A. M., Masalu, J. R. P., Mabelya, L., & Wandwi, W. B. C. (2009). Dental fluorosis in Tanzania Great Rift Valley in relation to fluoride levels in water and in 'Magadi' (Trona). *Desalination, 248*(1-3), 610–615. doi:10.1016/j.desal.2008.05.109.

Wacey, D., Saunders, M., Brasier, M. D., & Kilburn, M. R. (2011). Earliest microbially mediated pyrite oxidation in ~3.4 billion-year-old sediments. *Earth and Planetary Science Letters, 301*(1-2), 393–402. doi:10.1016/j.epsl.2010.11.025.

Wang, H. (2007). A review on process-related characteristics of pyrrhotite. *Mineral Processing and Extractive Metallurgy Review, 29*(1), 1–41. doi:10.1080/08827500601139523.

Wang, Y., Sun, X., Cao, Y., Xu, Z., Sun, H., Guan, X., Ghani, M. U., Zheng, L., Li, B., Huang, D., & Sun, W. (2025). Identification of hydrogen oxidation coupled with antimonate reduction, a novel antimony biogeochemical cycling, in two contrasting antimony-contaminated environments. *Microbiome, 13*(1). Available from https://microbiomejournal.biomedcentral.com/ . doi:10.1186/s40168-025-02149-z.

Wang, Z., Li, X., Song, J., Zhong, G., Liu, S., He, Z., Ma, J., Yuan, H., & Duan, L. (2025). A new indicator can assess absorption capacity for carbon dioxide and ocean acidification. *Communications Earth and Environment, 6*(1), 4–13. https://doi.org/10.1038/s43247-025-02380-4.

Wilkin, R. T., & Barnes, H. L. (1997). Formation processes of framboidal pyrite. *Geochimica et Cosmochimica Acta, 61*(2), 323–339. doi:10.1016/s0016-7037(96)00320-1.

Wilkin, R. T., Barnes, H. L., & Brantley, S. L. (1996). The size distribution of framboidal pyrite in modern sediments: An indicator of redox conditions. *Geochimica et Cosmochimica Acta, 60*(20), 3897–3912. doi:10.1016/0016-7037(96)00209-8.

Wodrich, M. D., Hall, S. J., & Zhu-Barker, X. (2026). Acidification and nitrification inhibition of manure alters greenhouse gas emissions and nitrogen cycling in diverse agricultural soils. *Journal of Environmental Quality, 55*(1). Available from https://acsess.onlinelibrary.wiley.com/loi/15372537 . doi:10.1002/jeq2.70123.

Wolkersdorfer, C., Nordstrom, D. K., Beckie, R. D., Cicerone, D. S., Elliot, T., Edraki, M., … Soler i Gil, A. (2020). Guidance for the Integrated Use of Hydrological, Geochemical, and Isotopic Tools in Mining Operations. *Mine Water and the Environment, 39*(2), 204–228. https://doi.org/10.1007/s10230-020-00666-x.

Worlanyo, A. S., & Jiangfeng, L. (2021). Evaluating the environmental and economic impact of mining for post-mined land restoration and land-use: A review. *Journal of Environmental Management, 279*. Available from https://www.sciencedirect.com/journal/journal-of-environmental-management . doi:10.1016/j.jenvman.2020.111623.

Xia, Z., Du, Z., Zhou, X., Jiang, S., Zhu, T., Wang, L., Chen, F., Carvalho, L., Zou, M., Becerra Lopez-Lavalle, L. A., Zhang, X., Xu, L., Wang, Z., Chen, M., Guo, X., Wang, S., Li, M., Li, Y., Wang, H., … Wang, W. (2025). Pan-genome and haplotype map of cassava cultivars and wild ancestors provide insights into its adaptive evolution and domestication. *Molecular Plant, 18*(6), 1047–1071. Available from https://www.sciencedirect.com/science/journal/16742052 . doi:10.1016/j.molp.2025.05.014.

Xiong, Q., Zhang, L., Xun, A., & Du, J. (2025). Review on the Application Research of Covers with Capillary Barrier Effects. *Journal of Research in Science and Engineering,* 7(1), 125–139. https://doi.org/10.53469/jrse.2025.07(01).19.

Yadollahi, A., Abdollahi, H., Dolati Ardejani, F., Mirmohammadi, M., & Magdouli, S. (2021). Bio-oxidation behavior of pyrite, marcasite, pyrrhotite, and arsenopyrite by sulfur- and iron-oxidizing acidophiles. *Bioresource Technology Reports, 15*, 100699. doi:10.1016/j.biteb.2021.100699.

Yao, W.-, Lei, H., Yang, Z.-, & Zhao, F.- (2022). Effects of organic acids on heavy metal release or immobilization in contaminated soil. *Transactions of Nonferrous Metals Society of China, 32*(4), 1277–1289. doi:10.1016/s1003-6326(22)65873-4.

Yeşiller, N., Hanson, J. L., Kopp, K. B., & Yee, E. H. (2016). Heat management strategies for MSW landfills. *Waste Management, 56*, 246–254. https://doi.org/10.1016/j.wasman.2016.07.011.

Yeşiller, N., Hanson, J. L., & Yee, E. H. (2015). Waste heat generation: A comprehensive review. *Waste Management, 42*, 166–179. https://doi.org/10.1016/j.wasman.2015.04.004.

Zawadzka-Małota, I. (2015). Testing of mining explosives with regard to the content of carbon oxides and nitrogen oxides in their detonation products. *Journal of Sustainable Mining, 14*(4), 173–178. https://doi.org/10.1016/j.jsm.2015.12.003.

Zhang, N., Xing, J., Wei, L., Liu, C., Zhao, W., Liu, Z., Wang, Y., Liu, E., Ren, X., Jia, Z., Wei, T., Siddique, K. H. M., & Zhang, P. (2025). The potential of biochar to mitigate soil acidification: a Global meta-analysis. *Biochar,* 7(1). Available from https://www.springer.com/journal/42773 . doi:10.1007/s42773-025-00451-5.

Zhang, W., Ozden, A., Yang, Y., Xu, A., & Li, F. (2025). Hidden acidification challenges in electrochemical ocean decarbonization. *ACS Energy Letters, 10*(6), 2788–2792. https://doi.org/10.1021/acsenergylett.5c00900.

Zuhra, K., & Szabo, C. (2022). The two faces of cyanide: An environmental toxin and a potential novel mammalian gasotransmitter. *The FEBS Journal, 289*(9), 2481–2515. doi:10.1111/febs.16135.

Zhang, A., Wilson, D., Ptacek, C. J., & Blowes, D. W. (2024). Reactive transport modelling of tailings hydrogeochemistry under a composite cover. *Journal of Contaminant Hydrology, 261*(June 2023), 104290. https://doi.org/10.1016/j.jconhyd.2023.104290

Zaharescu, D. G., Burghelea, C. I., Dontsova, K., Presler, J. K., Hunt, E. A., Domanik, K. J., Amistadi, M. K., Sandhaus, S., Munoz, E. N., Gaddis, E. E., Galey, M., Vaquera-Ibarra, M. O., Palacios-Menendez, M. A., Castrejón-Martinez, R., Roldán-Nicolau, E. C., Li, K., Maier, R. M., Reinhard, C. T., & Chorover, J. (2019). Ecosystem-bedrock interaction changes nutrient compartmentalization during early oxidative weathering. *Scientific Reports, 9*(1), 1–16. https://doi.org/10.1038/s41598-019-51274-x

CHAPTER TWO

Prediction of acid mine drainage

2.1 Predicting AMD

Predicting acid mine drainage (AMD) involves characterization of mine waste samples through all or either of the following: (1) static, (2) kinetic laboratory tests, (3) field laboratory tests, and (4) mathematical modelling to assess acid-generating potential versus neutralizing capacity (Plante et al. 2020; Skousen et al.,1997). In order to achieve these, proper sampling is key. Representative sampling is a crucial step influencing the subsequent conclusions that will be drawn from the characterization steps (Elghali et al., 2023; Parbhakar-Fox & Lottermoser, 2015). Sampling for AMD prediction is undertaken right from exploration to mine closure (Blannin et al., 2022; McLemore, Smith, & Russell, 2007). Inadequate sample selection and poor sampling techniques can contribute to excessive variance, difficulties in interpretation and incorrect assessment in an AMD prediction program. As a result of poor sampling, there may be serious consequences and costs (Downing, 2014). Thus the collected samples must be as representative as possible of all geological, lithological and alteration units related to the mine development plan and be representative of the relative amounts and particle size of each type of materials (Downing, 2014; SRK, 1989).

For representative samples' sake, Parbhakar-Fox and Lottermoser (2015) suggest that (1) geo-environmental ore-deposit models should be consulted at the exploration stage to understand potential outcomes and anticipated environmental consequences of mining for the deposit in question and that through consultation of these models, forecasting of potential rehabilitation requirements is permitted; (2) the number of samples required for ARD tests must be determined at the exploration stage—thus sufficient fund to support has to be apportioned. Plante et al. (2020) recommend to estimate the number of samples (N) using (SRK, 1989) rule of thumb which uses Eq. (2.1) where M is the mass (ton) of a given geological unit:

$$N = 0.026M^{0.5} \tag{2.1}$$

Elghali et al. (2023) refined the (Parbhakar-Fox & Lottermoser, 2015) proposed sampling guideline by suggesting the following to be considered

Acid Mine Drainage: Sustainable Mining Practices for Mitigating Environmental Impact
DOI: https://doi.org/10.1016/B978-0-443-45107-2.00005-8

for representative sampling: (1) the budget allocated for the study, (2) the prior knowledge of the study area, (3) the heterogeneity of the lot to be sampled in terms of its physical, chemical and mineralogical properties, and (4) the objective of the study. For estimation of error, Elghali et al. (2023) recommend to apply Pierre Gy's formula (e.g., Svensmark, 2021) as in Eq. (2.2) to calculate the fundamental error (FE), where *Ms* is the weight of the sample (g), *Ml* is the weight of the lot (g), corresponding to 95% passing on the particle size distribution curve, and *k* is defined as in Eq. (2.3) where *f* is the shape factor, *g* is the factor related to the particle size distribution, *c* is the mineralogical composition factor, and *l* is the liberation factor (varies between 0 and 1).

$$\sigma^2(FE) = K \times d^3 \times \left(\frac{1}{Ml} \times \frac{1}{Ms}\right) \tag{2.2}$$

$$\mathrm{k} = \mathrm{f} \times \mathrm{g} \times \mathrm{c} \times l \tag{2.3}$$

Following the sampling guidelines proposed by the aforementioned equations, authors hope that one can confidently propose the representative samples for AMD projects. Although ARD/AMD looks like characterization of waste rock and tailings (upstream), the sampling stage must emphases on a broad coverage to include the receiving environment (downstream), that is, surrounding soil, stream sediment and water if available which are critical for establishing baseline data necessary for sound ARD/AMD database as suggested by (Downing, 2014).

2.2 Short-term field tests or screening tests

Low-cost, on-site field tests are an essential part of any ARD test work program because they are crucial for selecting what and how much to be collected at the site for further analysis in the laboratory; thus the tools used must be quick, easy to use and inexpensive static tests that can help to establish as many as possible samples within a short time (Valcárcel, Cárdenas & Gallego, 1999). The most commonly pursued tests and practices include drill core observations, field leach tests (FLTs) for pH and dissolved metals and analyses for total metal concentrations using portable X-ray fluorescence instruments (Parbhakar-Fox & Lottermoser, 2015).

A drill core is cylindrical sample of a rock captured during drilling using a drill bit made up of steel and diamond (Baker et al., 2015). Drill core samples are commonly used for metallurgical, geotechnical and resource evaluation estimations, and they provide valuable interpretation of the nature and relationships between sulfide and carbonate mineralogy on a core scale but indirectly provide an efficient way to gather a deposit-scale understanding of the potential for acid formation (Parbhakar-Fox & Lottermoser, 2015). Logging generally provides a high-quality, undisturbed and chronologically ordered record of the underground geological formations/lithologies. Parbhakar-Fox et al. (2011) developed an ARD index (ARDI)—a site-specific logging code designed to numerically score five key parameters that are known to control ARD formation, including (1) sulfide and carbonate mineralogy, (2) primary and secondary neutralization mineralogy, (3) alteration variations, (4) modal mineralogy, and (5) grain boundaries as reaction sites for acid formation and neutralization.

Paste pH is a laboratory test that measures acidity or alkalinity of a sample, typically soil, rock or other materials, by mixing a solid with water to form a paste and then measure its pH with a pH meter (Olds et al., 2016; Sobek et al., 1978). Samples with paste pH values below 4.0 are considered potentially acid-forming (PAF) that will immediately produce acid upon exposure to water, while those with paste pH between 4.0 and 5.0 are also PAF, but with a lower stored acidic–salt content, suggesting a slight delay before significant acidity is produced (Weber, Hughes, Conner, Lindsay, & Smart, 2006). Samples with these low-pH values are also indicative of reactive sulfide minerals (Olds et al., 2016). Circumneutral paste pH values (approximately 6.5–7.5) suggest sufficient acid-neutralizing capacity to temporarily prevent acid generation, although the actual water quality values still needs to be confirmed through further testing (Goumih et al., 2022; Marmier et al., 2025b). Together with electrical conductivity, paste pH gives an indication of salinity of the waste material when initially exposed in a waste emplacement area (Smart et al., 2002). Paste pH is determined in pulverized samples (e.g., <75μm in size) mixed with deionized water at a ratio of 1:2 or 1:1 solid to water (Smart et al., 2002; Sobek et al., 1978). Although considered to be appropriate for only measuring present/current rather than future acid generation, paste pH tests remain a widely used, rapid and effective tool for assessing first flush characteristics in predictive ARD/AMD (Mugova & Wolkerdorfer, 2011; Zhang et al., 2026).

Other tests include "the field leach test." The United States geological survey FLT uses deionized water and a 5-min agitation period to quickly assess the readily soluble water reactive characteristics of a sample (Hageman, 2005). The FLT has expanded, and it is now used in diverse studies to characterize a wide range of other geological materials

(Hageman et al., 2015; Plumlee et al., 2013). Unlike paste pH, which uses saturated paste, FLT uses a 20:1 leaching ratio (water/solid), similar to other leaching tests such as USEPA methods 1311 and 1312 (USEPA, 1992, 1994). This ratio helps prevent solute saturation and provides enough leachate to complete all the desired analyses (Hageman et al., 2015).

2.3 Static tests

Static tests are the most widely used methods to predict acid generation potential of a mine waste because they are fast and relatively inexpensive initial assessment of the balance between acid-generating (e.g., sulfides) and acid-neutralizing (e.g., carbonates) components (Elghali et al., 2023; Karlsson et al., 2018). These short-term test methods for AMD are (1) paste pH, (2) chemical acid–base accounting (ABA), (3) mineralogical static test, and (4) net acid generation tests (NAG).

2.3.1 Acid–base accounting

An ABA test is a theoretical calculation of quantity of acid that will be produced if available sulfur in a sample is converted to sulfuric acid (Lawrence & Wang, 1997). An ABA test evaluates the balance between acid generation processes (e.g., oxidation of sulfide minerals) and acid neutralizing processes (e.g., dissolution of alkaline carbonates, displacement of exchangeable bases and weathering of silicates) (Lawrence & Wang, 1996; Paktunc, 1999a). The ABA test can be described as a difference between acid-producing potential (AP) or maximum potential acidity (MPA) and neutralizing potential (NP) or acid neutralization capacity (ANC) (Smart et al., 2002; Sobek et al., 1978). The difference between MPA and ANC is referred to as the net acid-producing potential (NAPP) and the units of NAPP are kgH_2SO_4/t as per (Eq. 2.4; Smart et al., 2002). Therefore, both MPA and ANC are expressed in kgH_2SO_4/t (Manchisi & Ndlovu, 2021; Stewart, Miller, & Smart, 2006). Alternatively, the net neutralization potential (NNP), which is the difference between NP and AP values with the units of kg $CaCO_3/t$, is expressed as per Eq. (2.5) (e.g., Price, 2009):

$$\mathrm{NAPP}\left(\frac{\mathrm{kgH_2SO_4}}{\mathrm{t}}\right) = \mathrm{MPA}\left(\frac{\mathrm{kgH_2SO_4}}{\mathrm{t}}\right) - \mathrm{ANC}\left(\frac{\mathrm{kgH_2SO_4}}{\mathrm{t}}\right) \quad (2.4)$$

$$\mathrm{NNP}\left(\frac{\mathrm{kgCaCO_3}}{\mathrm{t}}\right) = \mathrm{NP}\left(\frac{\mathrm{kgCaCO_3}}{\mathrm{t}}\right) - \mathrm{AP}\left(\frac{\mathrm{kgCaCO_3}}{\mathrm{t}}\right) \quad (2.5)$$

2.3.1.1 Determination of acid-producing potential

The AP, also referred to as MPA (see the previous section), is the total acid that mine wastes can produce from oxidation of their sulfur species content (Manchisi & Ndlovu, 2021). The principle behind determination of AP is that once the amount of total sulfur or sulfur species in a sample is determined in the laboratory, the values obtained are used to compute for the acid generation potential of the sample (Manchisi & Ndlovu, 2021). The amount of sulfur in the sample can be determined as total sulfur (in %), sulfide (S^{2-} or S_2^{2-}), sulfate (SO_4^{2-}) or organic sulfur. Assumptions made by Sobek et al. (1978) are (1) all sulfur occurs in pyrite, (2) the pyrite is completely oxidized to H_2SO_4 (Eq. 2.6; Karlsson et al., 2021) and (3) calcite neutralizes the produced acid as per (Eq. 2.7; Manchisi & Ndlovu, 2021). The overall equation for pyrite oxidation and acid neutralization by calcite is given in Eq. (2.8) (Chai et al., 2020; Dhir, 2018; Gomo, 2018):

$$FeS_2 + \frac{15}{4}O_2 + \frac{7}{2}H_2O \rightarrow Fe(OH)_3 + 2H_2SO_4 \tag{2.6}$$

$$2CaCO_3 + 2H_2SO_4 \rightarrow CaSO_4 + H_2O + CO_2 \tag{2.7}$$

$$FeS_2 + \frac{15}{4}O_2 + \frac{7}{2}H_2O + 2CaCO_3 \rightarrow Fe(OH)_3 + 2CaSO_4 + 2H_2O + 2CO_2 \tag{2.8}$$

Understanding the assumption of Sobek et al. (1978), reaction of Eqs. (2.6)–(2.8) and the units used (e.g. kgH_2SO_4/t or $kgCaCO\text{-}_3/t$) is very useful in understanding the mathematical expressions used in ABA test and other related tests such as kinetic tests. The MPA is mathematically expressed as in Eq. (2.8) (e.g., Schumann et al., 2012):

$$MPA\left(\frac{kgH_2SO_4}{t}\right) = (\%)S \times 30.6 \tag{2.9}$$

The 30.6 factor is based on the assumption that all sulfur measured by the total S method can be oxidized to produce 1 mol of sulfuric acid per mole of sulfur (Eq. 2.5). This stoichiometric factor, say x_1, can be calculated from molar mass ratio in Eq. (2.10) (Schumann et al., 2012):

$$X_1 = \frac{\text{Molar mass of } H_2SO_4}{\text{molar mass of S}} \times \left(\frac{1000\ kg/t}{100\%}\right) = \frac{98}{32} \times 10 = 30.6 \tag{2.10}$$

In the method of AP determination of Sobek et al. (1978), the total sulfur (% S) is determined using the combustion-iodometric titration method and the units used are kgCaCO/t. Acid production potential in terms of calcite equivalent is mathematically expressed in Eq. (2.11), where stoichiometric factor 31.25, call it x_2, can be calculated using Eq. (2.12) (e.g., Plante et al., 2020). The stoichiometric factor x_2 accounts for the amount of calcite that will be necessary to neutralize acid produced from oxidation of 1% of sulfur (pyrite) in 1 ton (or 1000 kg) of material based on assumption of Sobek et al. (1978) (Erickson & Hedin, 1988).

$$AP\left(\frac{\mathrm{kgCaCO_3}}{t}\right) = \mathrm{S}(\%) \times 31.25 \tag{2.11}$$

$$X_2 = \frac{\text{Molar mass of } \mathrm{CaCO_3}}{\text{Molar mass of S}} \times \left(\frac{1000\mathrm{kg/t}}{100\%}\right) = \frac{100}{32} \times 10 = 31.25 \tag{2.12}$$

2.3.1.2 Determination of neutralizing potential

To determine the neutralizing potential (NP), the sample is reacted with a known amount of standardized HCl and then back-titrated with standardized NaOH to determine the amount of HCl neutralized by the sample (Bouzahzah et al., 2015; Frostad et al., 2003; Skousen et al., 1997; Sobek et al., 1978). The determination of NP begins with a "fizz" test to obtain the sufficient amount of acid that will be added to a sample for total reaction of calcium carbonate present. One to two drops of 1:3 HCl are added to 0.5 g of the pulverized sample (<250 μm) to allow bubbling or audible Fizz, which indicates the presence of carbonate minerals. If no visible fizzing occurs, 20 mL of 0.1 N HCl is sufficient (Sobek et al., 1978). For a slight fizz, 40 mL of 0.1 N HCl is required; for a moderate fizz, 40 mL of 0.5 N HCl is used; and for a strong fizz, 80 mL of 0.5 N HCl is necessary to ensure complete reaction with all carbonate species. The suitable amount of HCl is then added to a 2.00 g sample, heated nearly to boiling temperature until gas evolution ceases and the particles settle; thereafter, water is added to 125 Ml to dilute the solution. The mixture is then boiled for 1 minute and slightly cooled above room temperature. It is then titrated with standardized NaOH until a constant reading of pH 7.0 remains for at least 30 seconds. The NP is calculated using Eq. (2.13). The factor 50 is the equivalent weight (g/equivalent) of $CaCO_3$ that converts mg equivalent weight to mg/g $CaCO_{-3}$ which is mathematically similar to kg/t $CaCO_3$

(Lawrence & Wang, 1996).

$$NP\left(kg\frac{CaCO_3}{t}\right) = \frac{(\text{HCl normality} \times \text{HCl volume(mL)}) - (\text{NaOH normality} \times \text{NaOH volume(mL)}) \times 50}{\text{sample weight (g)}} \tag{2.13}$$

Worked example 1

In a test to determine the NP of waste rock, 40 mL of 0.1N hydrochloric acid is added to 2.0 g of the sample. After the digestion, it is determined that 22.5 mL of 0.12 N sodium hydroxide is required to titrate the pulp to pH of 7 (Lawrence & Wang, 1996). Calculate the NP of the sample.

Solution

$$\text{HCl consumed/g} = \frac{(40\text{mL} \times 0.1\text{N of HCl}) - (22.5\text{mL} \times 0.12\text{N of NaOH})\text{mg equivalents}}{2\text{ g}} = 0.65\text{equivalent/g}$$

To convert the 50 g/equivalent (of HCl) to the units of NP that is kg$CaCO_3$/t, the equivalent weight can be calculated as per Eq. (2.14) (Davani et al., 2005) where *n* factor (valence factor) for $CaCO_3$ is 2 because $CaCO_3$ consumes 2 moles of H^+ from HCl as per Eq. (2.15) (Lawrence & Wang, 1996).

$$\text{Equivalent weight} = \frac{\text{molar mass of } CaCO_3}{n \text{ factor}} = \frac{100\text{ g/mol}}{n \text{ factor}} = 50\text{ g/equivalen} \tag{2.14}$$

$$NP = 0.65\frac{\text{equivalents}}{\text{g}} \times \frac{50\text{g}}{\text{equivalent}} = 32.5\text{kg } CaCO_3/t$$

The *n* factor (valence factor) for $CaCO_3$ is 2 because $CaCO_3$ consumes 2 moles of H^+ from HCl as per Eq. (2.15) (Lawrence & Wang, 1996).

$$CaCO_3 + 2HCL \rightarrow CaCl_2 + H_2O + CO_2 \tag{2.15}$$

2.3.1.3 Limitation of acid–base accounting test of Sobek et al. (1987)

Sobek et al. (1987) significantly contributed to development of standard ABA test which is a useful screening assessment tool for AMD with same modification. The method of Sobek et al. (1987) has certain limitations, which include (1) overestimation of AP due to presence of nonacid producing sulfur species in total sulfur, (2) potential for overestimation of NP due to presence of nonneutralizing carbonate minerals, (3) more silicate dissolution in laboratory conditions than in typical storage conditions due to heating and strong acids which overestimates the NP, and (4) the interpretation of a fizz-rating is subjective to practitioners, potentially resulting in the wrong strength of reagents later used in determination of NP (Karlsson et al., 2018; Manchisi & Ndlovu, 2021; Parbhakar-Fox & Lottermoser, 2015; Plante et al., 2020).

Overestimation of AP can occur because not all forms of sulfur species in the waste rock or tailings produce acid. Certain sulfides, some sulfates and organic sulfur do not generate acid (Manchisi & Ndlovu, 2021). For examples, while iron sulfide minerals (pyrite, pyrrhotite, arsenopyrite) are known to generate acidity chalcocite and covellite yield less acid than pyrite, sphalerite and galena may not yield acid at all (Jennings, Dollhopf & Inskeep, 2000). Moreover, while the sulfate–sulfur in gypsum ($CaSO_4.2H_2O$), anhydrite $CaSO_4$, and barite $BaSO_4$ minerals do not generate acid, the sulfates in jarosite ($AlFe_3(SO_4)_2.(OH)_6$), alunite ($KFe_3(SO_4)_2.(OH)_6$), and melanterite ($FeSO_4.7H_2O$) are acid-generating (Manchisi & Ndlovu, 2021). Water-soluble melanterite undergoes oxidation and hydrolysis in moist oxidative environments, while jarosite, which is stable at a pH <4 and relatively soluble above pH 4.7, undergoes hydrolysis in moist environments without the need for oxygen to release the stored acidity (Li et al., 2007).

Regarding NP, the presence of nonneutralizing carbonate minerals and enhanced dissolution of reactive silicate minerals in laboratory conditions causes overestimation of NP (Lawrence & Wang, 1997). Usually, not all carbonate materials are responsible for alkalinity or neutralization of acids. For example, in contrast to calcite and dolomite, siderite (or $FeCO_3$) is ineffective in acid neutralization because, although it initially consumes acid released, forming Fe^{2+}, the Fe^{2+} can oxidize to Fe^{3+} and hydrolyze, generating acidity that counterbalances the neutralization (Eqs. 2.16– 2.18). This results in zero net alkalinity by siderite (Skousen et al., 1997).

$$FeCO_3 + 2HCl \rightarrow Fe^{2+} + 2Cl^- + H_2O + CO_2 \tag{2.16}$$

$$Fe^{2+} + \frac{1}{4}O_2 + HCl \rightarrow Fe^{3+} + 2Cl^- + \frac{1}{2}H_2O \tag{2.17}$$

$$Fe^{3+} + 3NaOH \rightarrow Fe(OH)_3 + 3Na^+ \tag{2.18}$$

Because the net alkalinity by siderite's carbonate is zero, the NP calculated from a material containing the siderite is relatively lower, irrespective of the material's total carbonate (Skousen et al., 1997). When there is insufficient time for oxidation of Fe^{2+} to Fe^{3+} and subsequent precipitation of $Fe(OH)_3$, only the base-generating part of the reaction is counted; erroneously high NP values can be reported for samples containing siderite as a result (Skousen et al., 1997). Because of these limitations, the Sobek method has been accurate in some instances and misleading in others; thus research has continued on improving predictive methods (Erickson & Hedin, 1988).

2.3.2 Modifications of Sobek's ABA test

Numerous studies have been undertaken to enhance the reliability of ABA tests for AMD prediction in different approaches, including (1) consideration of sulfate salts such as jarosite, which produce acidity on hydrolysis to counteract the underestimation of AP, (2) computation of AP from sulfide–sulfur instead of total sulfur to counteract the overestimation of AP from total sulfur, (3) computation of NP from inorganic carbonate, and (4) consideration of mineral reactivity rates in computation of NP. Therefore, these modifications can be grouped into two main categories: (1) determination of carbonate species and (2) determination of sulfur species.

2.3.2.1 Determination of carbonate species

Lawrence and Wang (1997) modified the NP procedure of Sobek et al. (1978) by replacing the hot acid digestion with a controlled ambient-temperature HCl digestion carried out at 25–35°C for about 24 hours. Moreover, instead of relying on the fizz rating, acid addition is adjusted to maintain the final pulp pH at 2.0–2.5. The volume of HCl to be added is divided into three time periods at 0 h, volume (mL) added is 1, 2, 2 and 3 for none, slight, moderate and strong fizz rating, respectively, whereas at 2h time, the volume of 0.1N HCl added is 1, 1, 2 and 2 for no, slight, moderate and strong fizz rating, respectively. After 24 h, the acid addition is terminated and the distilled water is added to make a final volume of 125 mL, after which the solution is titrated against standardized 0.1 N NaOH to endpoint pH of 8.3 instead of 7; thereafter, NP is calculated same way as per original Sobek et al.'s (1978) test. The pH 8.3 is the stoichiometric end point of carbonate/bicarbonate in natural waters (Andersen, 2002).

Neutralization potential can also be calculated from % w/w carbon content using Eq. (2.19). Where the molar mass ratio M_{CaCO3}/M_C of calcite to carbon converts % carbon content to carbonate, 100 kg/t ÷ 100% accounts for 1% of carbon per ton of the waste rock (Lawrence & Wang, 1996)

$$\text{Carbonate NP}\left(\frac{\text{kgCaCO}_3}{\text{t}}\right) = \%\text{C} \times \frac{\text{M}_{\text{CaCO}_3}}{\text{M}_\text{c}} \times \frac{1000\ \text{kg/t}}{100\%} = \%\text{C} \times 83.3 \tag{2.19}$$

In rocks, soils and mining wastes, carbon occurs in various forms that can be broadly categorized into two main groups: (1) inorganic carbon (C_{inorg}) and (2) organic carbon (C_{org}) (Shamrikova et al., 2024; Stojanova et al., 2024). Inorganic carbon is primarily present as carbonate minerals, most of which are acid-neutralizing, although some (e.g., siderite) may not contribute significantly to acid neutralization (Stewart, Miller, & Smart, 2006). Methods for determining these carbon forms are described in detail by Shamrikova et al. (2024) and Stojanova et al. (2024).

Total carbonate concentration can be quantified by high temperature combustion at 1500°C–1700°C in presence of O_2 (Klingshirn et al., 2019; Peterson et al., 2003; Watanabe et al., 2007). At this temperature, carbon is converted to CO_2 and the evolved CO_2 can be quantitatively measured by infrared spectrometry, standard volumetric or titrimetric methods (Bradshaw & Brewer, 1988; Lapakko, 2002; Turek et al., 1978). Alternatively, coulometric methods can be used to determine carbonate-carbon (Engleman, Jackson & Norton, 1985). This method is time costing and is therefore rarely used in AMD. Thereafter, organic carbon can be determined using (1) pyrolysis or (2) acid digestion method (Engleman, Jackson & Norton, 1985).

2.3.2.1.1 Acid digestion method

This method assumes that (1) acid digestion dissolves all carbonates present and (2) residual solid contains only organic carbon and graphite. The analysis procedure involves the digestion of sample with hot 20% HCl, dry and rinse residual chloride, which can interfere with the subsequent analysis for total carbon (Kieber et al., 2002; Wangersky, 1993). The residual solid is analyzed for total carbon and assumed to be organic carbon (Lapakko, 2002). The carbonate carbon content is the difference between total carbon and acid insoluble carbon. Other acids such as HCl_4, HNO_3, H_2SO_4, and H_3PO_4 can be used for removing inorganic carbonate, but the limitations associated must be addressed (Bisutti, Hilke & Raessler, 2004). For example, HNO_3 and H_2SO_4 are strong oxidizing agents associated with high loss of organic carbon up to 80% according to Bisutti, Hilke & Raessler (2004). H_2SO_3 removes carbonate except siderite and dolomite and does not react with labile organic carbon (Bisutti et al., 2004; Gibbs, 1977); failure to remove dolomite may underestimate NP calculated from net carbonate considering that dolomite is acid neutralizing (Chai et al., 2020). $HClO_{-4}$ reacts quantitatively with inorganic carbon and does not show any oxidation power at room temperature in aqueous solutions, but during the drying of the sample after acid treatment, it forms perchlorates that are hazardous when heated (Fabin & Jarosz, 2025).

2.3.2.1.2 Acetic acid soluble calcium

Acetic acid selectively dissolves carbonate minerals, especially calcite and dolomite as per Eq. (2.20) while minimizing the dissolution of silicate minerals (Kim & Yoon, 2021). The sample is subjected to digestion in 10% acetic acid at 90°C for 30 minutes, an aliquot of the solution is taken and diluted to volume, the sample is preserved with 2% HCl, lanthanum (1%) is added to inhibit matrix effects and the solution is analyzed for calcium using spectroscopic technique such as FAAS (Lapakko, 2002). The obtained Ca^{2+} can be converted to carbonate equivalent and be used in calculation of NP (Kim & Yoon, 2021).

$$CaCO_3 + 2CH_3COOH \leftrightarrow Ca(CH_3COO)_2 + H_2O + CO_2 \tag{2.20}$$

2.3.2.1.3 Pyrolysis method

Pyrolysis method involves heating samples at 550°C for one hour to drive off organic carbon as carbon dioxide (and sulfide sulfur as sulfur dioxide) after which the carbonate carbon is estimated as the total carbon in the residue. Although this method has advantage of determining both carbonate carbon and sulfide–sulfur contents with the same preparatory step, underestimation of carbonate carbon is likely due to loss of some carbonate during pyrolysis (Lapakko, 2002); dehydration of structural water at higher temperature may be misinterpreted as organic matter loss leading to overestimation of organic carbon content (Giraldo et al., 2024)

2.3.2.2 Determination of sulfur species

Sulfur speciation is crucial for understanding AMD because different sulfur species have varying potentials to generate acidity. The categories of sulfur species in geological materials are broadly divided into (1) sulfide, (2) sulfate, (3) organic sulfur, and (4) elemental sulfur (Akinwekomi et al., 2020; Casagrande, Finkelman and Caruccio, 1989; Czerewko et al., 2003; León et al., 2024). These distinct forms of sulfur can be analyzed separately within the rock samples (Czerewko et al., 2003). This subsection is divided into two parts; the first part describes various techniques used in analysis of sulfur in geological materials with emphasis on advancement in technology; the second part discusses techniques used for analysis of sulfur species focusing on sulfide for AP or MPA are discussed.

2.3.2.3 Analytical techniques for sulfur in geological materials

In rock samples or any geological material, total sulfur can be determined by several methods, including barium sulfate gravimetric, combustion-volume, titration colorimetric, ion chromatography, XRF, ICP-OES, and ICP-MS (e.g., Zhang et al., 2024). Each method requires sample preparation to expose the analyte that can be detected and quantified. These are (1) pressing and melting technique, (2) high temperature combustion, (3) high pressure combustion in oxygen-rich environment, and (4) acid digestion. Pressing (pelletization) and melting techniques are typically used for XRF analysis. Pressing involves compressing the powders directly into a stable disc with or without a binder to form pallets that are placed on XRF for analysis (e.g., Kanrar, Sanyal & Dhara, 2021). Melting (or fusion) involves powder mixed with fluxes and melted into a glass disc (e.g., Kanrar, Sanyal & Dhara, 2021). The XRF analysis is fast, nondestructive, and reproducible and requires minimal chemical handling. However, its use for total sulfur determinations remains only for rapid screening because of limitations, including a higher detection limit and matrix interference, which cannot be eliminated by simple pressing (Zhao et al., 2020).

Another sample preparation technique is the combustion method. In this method, the sample is combusted in an oxygen-rich environment, either under high pressure within a bomb combustion system, at high temperature in a furnace or in a simple sealed Schöniger apparatus (Azevedo et al., 2025; Coller and Leininger, 1955; Hubbard, Katz and Waddington, 1954). Combustion converts all sulfur into SO_2, which can be measured directly by infrared (IR) absorption, thermal conductivity or coulometric titration (Coller & Lelnlinger, 1955). Alternatively, in bomb combustion, the SO_2 is absorbed into a solution (distilled water or H_2O_2) and converted to H_2SO_4 (Hubbard, Katz & Waddington, 1954). The resulting SO_4^{2-} is then quantified either gravimetrically as $BaSO_4$, titrimetrically or by ion chromatography (Bozkaya & Aluç, 2024). Advantages of the combustion method for total sulfur determination include providing an accurate measure of all sulfur forms with minimal reagent use and reduced contamination risk. Disadvantages are also present—the method requires specialized equipment (e.g. bomb combustion), high-temperature or high-pressure conditions and careful handling due to safety risks.

Total sulfur can also be determined by spectroscopic techniques such as ICP-OES. ICP-OES works with samples in aqueous form and therefore sample digestion is critical. Heating the samples in mixture of acids and in a closed system is the recommended method that increases effectiveness while minimizes cross-contamination and loss of sulfur by volatilization (Robinson, 2025). For example, Zhang et al. (2024) suggest that digesting the sample at 180°C in anti-corrosion acid digestion bombs using a mixture of aqua regia ($HNO_3 + HCl$), and hydrofluoric acid (HF) could accurately quantify sulfur in high sample content material. The HF enhances dissolution by breaking down silicates, whereas aqua regia oxidizes S2– completely to S and then to SO_4^{2-} (Ezz-Eldin, Abd-Elaziz & Elalaily, 2010). Limitations of ICP based techniques for total sulfur determination in geological material are (1) the use of highly hazourdous acids such as HF and (2) spectral interference on sulfur lines by Ca and Al (e.g., Ackerman, Rohovec & Šebek, 2012).

2.3.2.4 *Determination of sulfur species for AP and MPA*

Understanding all sulfur forms in a sample is essential in assessing AMD because sulfide minerals (e.g., pyrite and pyrrhotite) are the major causes of AMD while most of sulfates and organic sulfur are not acid generating (Li et al., 2007; Belzile et al., 2004). Using total sulfur can therefore overestimate amount of AMD. Thus it recommends assessing AP using sulfide sulfur instead of total sulfur because a total sulfur approach does not take non-AP sulfates (e.g., gypsum and barite) into consideration and so can overestimate the AP of a sample. They propose that if sample is known to contain a significant barite content, barite correction can be done by first analyzing for barium and calculating the stoichiometric equivalent of sulfur and then sulfide sulfur is given by Eq. (2.21). Thereafter AP can be calculated using Eq. (2.22) instead of total sulfur. Works by Smart et al. (2002), Bouzahzah et al. (2015) and others reiterate the calculation of AP from sulfide instead of total sulfur, where sulfide can be determined directly like CRS method or indirect by subtracting sulfate sulfur from total sulfur (Black et al., 2018).

$$\text{Sulfide} = (S_{total} - S_{sulfate})\% \tag{2.21}$$

$$\left(\text{kgCaCO}_3\right)/\text{t}) = \text{Sulfide}(\%) \times 31.25 \tag{2.22}$$

Fractionation of sulfur species to obtain sulfide–sulfur can be achieved by both combustion and acid digestion methods (Laban & Atkin, 2000). This can be achieved by adjusting the sample pretreatment conditions, whether using combustion or acid digestion.

2.3.2.4.1 Combustion method

At higher temperatures (e.g., 1450°C), all forms of sulfur will be converted to SO_2 while at temperatures of 550–650°C in presence of O_2, sulfide–sulfur is volatilized and lost as gas while it will remain as residue (Karlsson et al., 2022). The total sulfur content in the pyrolysis residue is then determined. According to Karlsson et al. (2022), sulfide–sulfur is calculated from the difference between total sulfur in the untreated sample and total sulfur in the pyrolysis residue.

2.3.2.4.2 Acid digestion

In acid digestion, sulfur species are fractionated based on their solubility and reactivity with specific acids (Laban & Atkin, 2000; Claff et al., 2010). Sulfate–sulfur is extracted using HCl and quantified by gravimetric determination of a barium sulfate precipitate. Pyritic sulfur can be extracted using HNO_3 as a strong oxidizing agent, which dissolves and oxidizes the pyrite into soluble sulfate species (Karaca et al., 2003; Zárate-gutiérrez et al., 2012).

Sequential dissolution method is commonly employed to determine sulfur species in rock samples by selectively extracting forms of sulfur in the solution through stepwise chemical treatments (Matsumoto et al., 2018). Laban and Atkin (2000) proposed a method for the direct determination of sulfate–sulfur, pyritic sulfur, and organic sulfur by ICP-AES using digestants obtained through a sequential dissolution procedure with microwave digestion. The procedure involves three stages: (1) sulfate–sulfur was extracted by digesting 5 g of coal sample with 10 mL of 5 M HCl in a microwave system programed at 100% power for 5 minutes, followed by 50% power for 5 minutes. The residue from this step was washed with deionized water to remove residual HCl. (2) Pyritic sulfur was determined by digesting the residue with 10 mL of 2 M HNO_3 in the microwave, under the same program as the first stage. (3) Organic sulfur was determined by completely digesting the residue from stage two using concentrated HNO_3, HCl, HF, and boric acid. Results show that the sums of sulfate–sulfur, pyritic sulfur, and organic sulfur from sequential extraction agreed well with certified total sulfur values with good reproducibility, most sulfur forms having a coefficient of variance <15%.

2.3.2.4.3 The chromium reducible sulfur method

The chromium reducible sulfur (CRS) method was developed by Zhabina and Volkov (1987) for the speciation of sulfur in sediment samples and has since been explored for various other applications (Cornwell & Morse, 1987;

Morse & Cornwell, 1987; Burton et al., 2008). In Australia for instance, the CRS has been developed for acid sulfate soils and coal and is considered the suitable technique for mine wastes with low sulfur content, especially absence of pyrite and arsenopyrite (Black et al., 2018). This analysis involves converting reduced inorganic sulfur to H_2S using a hot acidic $CrCl_2$ solution. The released H_2S is captured in a zinc acetate solution as ZnS, which can be quantified through iodometric titration (Pawlak & Pawlak, 1999; Murcia et al., 2017). Alternatively, H_2S can be measured directly with a specific gas sensor (Yang et al., 1997). The CRS specifically targets inorganic sulfur with an oxidation state below +6 (Burton et al., 2008). The reactions may involve straightforward dissolution of monosulfides in acid to produce H_2S, reduction of disulfide sulfur–sulfur bonds (such as in pyrite) to release H_2S, or a combination of both mechanisms place (Evangelou & Zhang, 1995; Narasagoudar et al., 1982; Wilkin & Barnes, 1996).

Schumann et al. (2012) investigated the utility of the CRS method for calculation of MPA, by determining recoveries of both acid producing (jarosite and melanterite) and nonacid forming (gypsum, barite, etc.) sulfate salts and organic sulfur compounds using CRS method as well as comparing the NAPP calculated from total sulfur, CRS (Eq. 2.23), and sum of CRS and jarosite (Eq. 2.24) with NAG and column leach tests for base metal mine wastes (Fe, Cu, Ni, Zn, and Pb sulfides) and coal wastes. Their results reveal that CRS was useful for the assessment of coal wastes with high organic sulfur, but the technique underestimates the acid potential of weathered wastes. In the study of Schumann et al. (2012), NAPP calculated from CRS agree well with NAG than that of total S while NAPP calculated from sum of % CRS+% jarusite result in overestimation. Black et al. (2018) report similar results that CRS agrees with NAG tests except high sulfur minerals such as arsenopyrite and molybdenite which both gave no to very low (1%) recovery by CRS, respectively, compared to total oxidizable sulfur.

$$MPA = \%CRS \times 30.6 \tag{2.23}$$

$$MPA = (\%CRS + {}^{3}/_{3}\%S_{jarosite}) \times 30.6(\%) \tag{2.24}$$

2.3.3 Net acid generation test

The NAG test involves reaction of a sample with H_2O_2 to rapidly oxidize any sulfide minerals contained within a sample (Baek et al., 2021;

Karlsson et al., 2018; Matsumoto et al., 2018). The NAG test assesses measured acidity produced from H_2O_2 oxidation of mine wastes to correct for overestimation resulting from organic acids (Oh et al., 2017). During the NAG test, acid generation and acid neutralization reactions can occur simultaneously (Sandan & Nurcholis, 2024; Sapsford et al., 2008). Therefore, the end result represents a direct measurement of the net amount of acid generated by the sample expressed as kg H_2SO_4/ton. The sample is reacted with H_2O_2 overnight, and in the next day, the mixture is gently heated to accelerate any oxidation of any residual sulfide. Thereafter the mixture is vigorously boiled to decompose any residual H_2O_2. The mixture is allowed to cool, diluted to a desired volume (say 250 ml). At this point, a pH called NAGpH is recorded. The resulting liquor is titrated with standardized NaOH until pH 4.5 is reached; then titration continues to pH 7 (Sapsford et al., 2008). The concentration of NaOH to be used for titration is dictated by the NAGpH. That is, 0.10-M NaOH is used when NAGpH >2 and 0.50-M NaOH is used when NAGpH = 2. Titration to pH 4.5 accounts for acidity due to Fe, Al, and most of the hydrogen ions while any additional acidity accounted for intitration between pH 4.5 and pH 7 is usually indicative of soluble metals such as Cu and Zn (Smart et al., 2002). The NAG in value (kg H_2SO_4/ton) is calculated from the volume V (mL) and concentration M (moles/liter) of NaOH used in titration and weight W of sample used as per Eq. (2.25) (Manchisi & Ndlovu, 2021).

$$\text{NAG} = (49 \times V \times M) \div W \tag{2.25}$$

2.3.4 Leaching tests

Leaching tests are laboratory procedures that simulate the release of contaminants from solid materials (e.g. soil, mine waste) into a surrounding liquid medium (usually water) to evaluate the potential environmental impact and assess risks to groundwater and human health (Gao et al., 2025). The most commonly used leaching tests are (1) synthetic precipitation leaching procedure (SPLP) and (2) toxicity characteristic leaching procedure (Hageman et al., 2015; Parbhakar-Fox and Lottermoser, 2015).

USEPA Method 1312—a SPLP (USEPA, 1994)—has been used both in regulatory and more commonly in nonregulatory leaching studies. The SPLP is used to characterize water-soluble, water-reactive constituents that are released from samples that have been subject to agitation for 18 h. The standard SPLP requires crushing the samples to pass a 9.5-mm sieve and requires a 20:1 (leachant/solid) leaching ratio. It calls for the use of slightly acidified extraction fluids that were designed to simulate acid rain (USEPA, 1994). USEPA Method 1311 Toxicity Characteristic Leaching Procedure uses same particle size and solid-to-liquid ratio as SPLP (USEPA, 1992). After leaching process is completed, the leachate is collected and analyzed for water quality parameters including pH, electrical conductivity, anions such as sulfates, major, minor, trace elements, and many others depending on the study objective (Hageman et al., 2015).

In conclusion, the various chemical static methods reviewed in this chapter serve as essential screening tools that provide a rapid initial assessment of mine wastes, though they necessitate follow-up with advanced mineralogical analysis to precisely identify specific acid-generating and neutralizing sources. Because there is no single test that can be applied universally to all geological materials, selection must remain site specific, utilizing desktop analysis and short-term field data to determine the most appropriate ABA method for materials ranging from coal and metalliferous wastes to tailings and fresh rock. Despite their utility, chemical static tests face some limitations: (1) high interpretation variability among the chemical ABA methods as summarized in Table 2.1, (2) uncertain classification especially for low sulfidic mine wastes, (3) an inability to predict the temporal onset or lag time of AMD, and (4) a lack of specificity regarding which individual minerals contribute most to the overall acid–base balance. These constraints highlight the need for advanced approach for waste characterization including precise mineralogy, kinetic tests and modeling which are addressed in the following sections.

2.3.5 Mineralogy static tests

Contrary to chemical static tests that are generally based on elemental analysis of sulfur and carbon, the mineralogical static tests require computation for AP and/or NP based on the mineralogical composition of mine wastes as summarized in Table 2.2and detailed hereafter.

Kwong (1993) suggested a net acid production parameter M analogous to the NNP. The mineralogical index M can be calculated as indicated in Eq. (2.26). The Kwong (1993) method consist of two terms. The first term $(2 \times (1 - r + p))$ accounts for acid-producing potential and the second one $(2 \sum_{n=1}^{6} X_{mn} W_{mn})$ is similar to the neutralization potential. Unlike chemical static tests, the net acid production index is not expressed in kg $CaCO_{-3}/t$; instead, a mine waste is classified as acid generating if $M > 0$ and not acid

Table 2.1 Classification of mine wastes based on drainage pH, ABA test and NAG test.

Method	Acid producing	Uncertainty	Nonacid-producing	References
Paste pH	pH < 5	5 < pH > 10.0 (near neutral)	pH >10	(Manchisi & Ndlovu, 2021)
NPR = NP/AP	NPR < 1	1 < NPR < 3	NPR >3	(Adam et al. 1997)
Units of AP and NP are kg $CaCO_3$/t	NPR < 1	1 < NPR < 1.3–4	NPR >1.3–4	(Adam et al., 1997; Morin & Hutt, 1994)
	NPR < 1	1 < NPR < 2.5	NPR >2.5	(Adam et al., 1997; Botero et al., 2024)
NNP = NP–AP (kg $CaCO_3$/t)	NPR < 0	−20 < NNP < 20	NPR >0	(Botero et al., 2024; USEPA, 1994)
NAPP = MPA–ANC (kg H_2SO_4/t)	NAPP > 0	NR	NAPP<0	(Smart et al., 2002)
ANC/MPA	ANC/MPA < 1	NR	ANC/MPA>2	(Smart et al., 2002)
NAGpH and NAG value	NAGpH < 4.5 and NAG = 0	NAGpH = 4.5 and NAG = 5	NAGpH = 4.5 and NAG > 0	(Smart et al., 2002)
NAPP = MPA–ANC (kg H_2SO_4/t)	NAGpH < 4.5 and NAPP > 0	NAPP > 0 and NAGpH > 4.5 or NAPP < 0 and NAGpH =4.5	NAPP < 0 and NAGpH = 4.5	(Smart et al., 2002)

AMD, Acid mine drainage; *ANC*, acid neutralization capacity; *AP*, acid-producing potential; *MPA*, maximum potential acidity; *NAG*, net acid generation; *NAPP*, net acid-producing potential; *NP*, neutralizing potential; *NR*, not reported; NPR, neutralizing potential ratio

Table 2.2 Mineralogical static test methods for assessment of AMD/ARD.

Method	Definitions	Characteristics
Kwong (1993) $M = 2x(1 - r + p) - 2\sum_{n=1}^{3} X_{mn} W_n$ 2.26	M = Mineralogical net acid production index (M) X = molar proportion of sulfide minerals in the mine waste, p = proportion of ferrous iron in the metal sulfide associations, r = average metal/sulfur atomic ratio in the primary sulfide assemblage, n = number of mineral weathering classes, X_{mn} and W_n = fraction and weathering rate of each mineral class respectively	Takes carbonates and silicates into account for NP calculation. If M > 0 the sample is acid generating otherwise, it is nonacid generating. The M is expressed neither in kg $CaCO_3$/t nor kgH_2SO_4/t, hence not comparable to other static tests. It does not consider the Fe content within the neutralizing minerals
Lapakko (1994) $NP = (10 \times \%_{cal.}) + (11.9 \times \%_{mag.})$ 2.27	$\%_{cal}$ = calcite abundance (wt.%) in the mine waste $\%_{mag}$ = magnesite abundances (wt.%) in the mine waste	It takes only calcite and magnesite for NP calculation. It does not consider the contribution of silicates and dolomite on the NP.
Lawrence and Scheske (1997) $NP = 1000\frac{kg}{t} \times M_{CaCO_3} \times \sum_{i=1}^{n} \frac{C_{Mi} R_i}{M_{Mi}}$ .... 2.28 Units of NP are kg/$CaCO_3$/t	C_{Mi} = neutralizing mineral content (wt.%) R_i = reactivity factor of the neutralizing minerals M_{Mi} = molar mass of the neutralizing minerals M_{CaCO_3} = Molar mass of calcite	Takes neutralizing silicates into account, based on relative reactivity. Does not take into account the Fe content of neutralizing minerals, which diminishes the NP when released.

(continued on next page)

Table 2.2 Mineralogical static test methods for assessment of AMD/ARD—cont'd

Method	Definitions	Characteristics
Paktunc (1999) $NP = \sum_{i=1}^{k} \frac{98 \times 10 \times X_i \times C_i \times n_s}{n_i \times \omega_i}$ ....2.29 $AP = \sum_{s=1}^{m} \frac{n_s \times 98 \times X_s \times 10}{\omega_s}$ ....2.30 NP and AP units are kgH_2SO_4/t	98 = molecular weight (g/mol) of H_2SO_4 10 = conversion factor for recasting in kg/t (i.e., 1000 kg/t ÷ 100%), X_i = amount of mineral, i in wt.%, X_s = the amount of sulfide mineral, s in wt.%, C_i = number of nonoxidizable cations in one formula unit of neutralizing mineral i, n_s = moles of sulfuric acid formed by the oxidation of one mole of sulfide mineral s, n_i = moles of mineral required to consume ns moles of sulfuric acid produced by the oxidation of one mole of sulfide mineral s, ω_i = molecular weight of neutralizing mineral i (g/mol) ω_s = molecular weight of sulfide mineral s (g/mol) k = number of neutralizing minerals in the sample m = number of sulfide minerals in the sample	Takes only carbonates into account. Takes into account the content of nonoxidizable cations such as Fe, which diminishes the NP when released.
Modified Lawrence and Scheske (1997) by Plante et al. (2012)NP = $1000\frac{kg}{t} \times M_{CaCO_3} \times \sum_{i=1}^{n} \frac{C_{Mi} R_{iC_i}}{M_{Mi}}$ 2.31	where C_i is the sum of the stoichiometric coefficients of nonoxidizable/nonhydrolyzable cations in the carbonates and the other parameters are the same as for Eq. 2.28	This method combines characteristics of the Lawrence–Scheske and Paktunc methods, accounting for the silicate contribution to the overall NP and the nonoxidizable cations content of the Fe–Mn-bearing carbonates.

Source: From Plante, Bussière and Benzaazoua (2012)and Bouzahzah et al. (2014).
AP, Acid-producing potential; *NP*, neutralizing potential.

generating if $M < 0$. The methods take carbonates and silicates into account for NP calculation contrary to chemical tests such as NAG which may underestimate silicate contribution to overall NP due to their slow reaction. However, the Kwong (1993) method still has some limitations: (1) it requires a precise mineralogical characterization of the mine waste for which the conventional mineralogical techniques such as X-ray diffraction (XRD) and optical microscopy are not suitable, and (2) it does not consider the Fe content within the neutralizing minerals (e.g., siderite; e.g., Elghali et al., 2023).

Lapakko (1994) suggested another mineralogical method for NP calculation (Eq. 2.27) which assumes that only calcite and magnesite are the responsible for neutralizing acids. The main limitation of this method is the fact that it does not consider the contribution of silicates and dolomite on the NP of a mine waste, which may lead to an underestimation of the NP. Noting this, Lawrence and Scheske (1997) proposed NP determination by using relative mineral abundances (wt.%) in the mine waste sample taking into account neutralizing mineral content (wt.%), the reactivity factor of the neutralizing minerals including silicates (Eq. 2.28). The main limitation of Lawrence and Scheske (1997) is that it doesn't take into account the Fe content of neutralizing minerals (e.g., siderite), which diminishes the NP when released (Elghali et al., 2023; Bouzahzah et al., 2013).

Lawrence and Scheske (1997) were highly criticized by Paktunc (1999b) questioning (1) the validity of the use of a normative classification scheme known as the CIPW for determining mineral abundance in samples, especially in cases of secondary minerals, sedimentary mineral assemblages, and minerals formed during metamorphism and hydrothermal alteration; (2) the use of the calculation of weighted NP values using relative reactivities that were later on found to be overestimated when compared to the reactivities of actual samples in the laboratory; and (3) the fact that it does not consider the effect of Fe carbonates on the neutralization potential of the mine waste.

Paktunc (1999a) proposed a new approach to determine not only NP but also AP based on mineralogical composition (Eqs. 2.29 and 2.30) in Table 2.2. For NP, the Paktunc (1999a) method provides an unambiguous NP value for each mineral that is deemed to have a neutralization potential and a sum of which gives bulk NP value for the sample (Eq. 2.29). In this approach, NP of each mineral is calculated based on its abundance (wt.%) in the sample and the stoichiometry of neutralization reactions. The factor Ci that accounts for the number of nonoxidizable cations in one formula unit of neutralizing mineral *i* was also introduced (Paktunc, 1999c). Similarly, Paktunc (1999a) suggested that AP should be calculated using Eq. (2.30) to take into consideration the presence of more than one type of sulfide mineral in samples. The main limitation of the formula proposed by Paktunc (1999a) is the lack of the reactivity factor for the different sulfide minerals. To overcome this limitation, Bouzahzah (2013) suggested multiplying the AP of Paktunc (1999a) by taking into account the relative rates of acidity production of each sulfide by using a relative reactivity factor. This factor is determined by the average of the total acidity produced by each sulfide calculated relatively to the acidity production of pyrite (Chopard et al., 2015).

To account for the main limitations of the method of Lawrence and Scheske (1997) of not considering the effect of Fe-carbonates on the neutralization potential of the mine waste, Plante, Bussière & Benzaazoua (2012) suggested a new formula to consider the Fe content of neutralizing minerals. The authors introduced a new parameter "C*i*" to avoid overestimation of NP due to the presence of Fe carbonates in the mine wastes—this parameter was previously introduced by Paktunc (1999b). Thus the modified method of Lawrence and Scheske (1997) that combines the desired characteristics of Lawrence–Scheske and Paktunc methods, accounting for the silicate contribution to the overall NP and the nonoxidizable cations content of the Fe–Mn-bearing carbonates, was proposed (Eq. 2.31).

Mineralogical static tests still have limitations which are (1) mineralogical ABA methods—for instance, Paktunc (1999) requires detailed quantitative mineralogical data obtained from advanced techniques which are expensive and time-consuming compared to chemical ABA tests (Lawrence & Wang, 1997; Sobek et al., 1978); (2) the method involves complex stoichiometric calculations that account for variations in mineral composition and reaction pathways; for example, it is necessary to distinguish between pyrite oxidation and pyrrhotite oxidation reactions (Jamieson, Walker & Parsons, 2015); and (3) mineralogical static tests alone cannot predict the onset or lag time of AMD.

2.4 Kinetic tests

Kinetic tests are complementary to static tests undertaken to overcome discrepancies between conditions for short-term laboratory testing on crushed, sieved samples, and long-term behavior of coarser waste rock(s) observed in the field (Tabelin et al., 2022). Interactions between minerals and neutralization mechanism of materials being tested depend on time and environmental conditions which cannot be detected by static tests. Thus kinetic tests which can be run under laboratory and field conditions involve a number of measurements over time and are used to assess a range of AMD parameters that include sulfide reactivity, oxidation kinetics, lag time (if it exists), metal solubility, and the leaching behavior of test materials (Tabelin et al., 2022; Koc et al., 2023). A number of kinetic tests have been developed to complement static tests data in the prediction of AMD formation. These kinetic tests are broadly categorized into two main groups are (1) laboratory-based kinetic tests which includes humidity cell tests (HCT), leach columns, shake flasks, biokinetic tests, soxhlet extraction, and BC research confirmation tests, and (2) field-based kinetic tests such as the case with large-scale field test pads (i.e., bins or cribs or piles), mine wall stations, and routine site monitoring (Manchisi & Ndlovu, 2021). The most common kinetic test methods used for AMD include HCT, kinetic column leach (KCL) tests, and field tests (Davis et al., 2019; Moyo et al., 2025; Neupane and Donahoe, 2013; Sapsford et al., 2009). The laboratory kinetic test is performed to simulate natural weathering of mine wastes at laboratory scale using water and/or air to accelerate weathering of mine waste targeting oxidation of sulfide minerals by oxygen and subsequent carbonate dissolution. Specific procedures on how to perform laboratory kinetic tests have been described in several text books and review papers (Elghali et al., 2023; Koc et al., 2023; Manchisi & Ndlovu, 2021; Parbhakar-Fox & Lottermoser, 2015; Vriens et al., 2020).

2.4.1 Kinetic column leach and humidity cell test

In KCL tests such as the EPA method 1627 (Fig. 2.1), about 1.8–2.0 kg of a reconstructed sample (with known particle size distribution) is fed into 2.5-feet high and 2 in diameter column into which the weekly leaching cycle with 6 days aeration using a humidified air at constant temperature and flow rate of 1 L/min and in a 7th day the column is saturated with deionized water and then drained and leachate samples collected for analysis. The analyzed parameters are pH, specific conductance, alkalinity, acidity, leachate volume, major elements (Ca, Mg, Ka, Na, SO_4), trace/minor elements (Al, Fe, Mn, Zn and Se), gypsum saturation index, calcite saturation index, temperature,

Figure 2.1 Example of kinetic leach column: (A) the geometry and cross-section of column and (B) a view from columns. *(From Erguler et al. (2014)).*

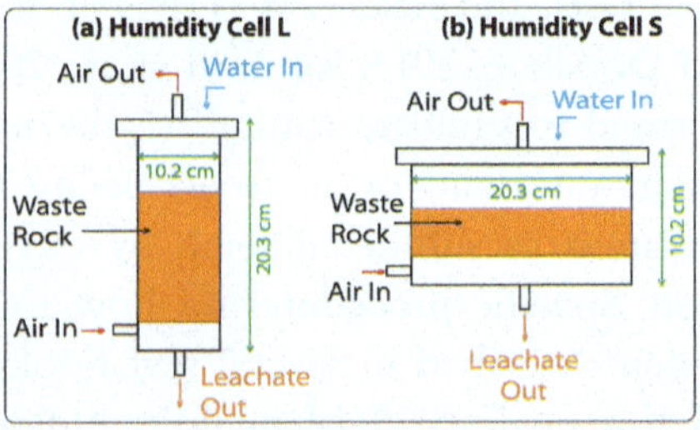

Figure 2.2 Schematic of two humidity cells, with distinct shapes and dimensions: (A) long and narrow humidity cell and (B) short and wide humidity cell *(From Pieretti et al. (2022)).*

gas inlate flow rate and % CO_2 in the discharge air on daily basis (USEPA, 2011).

The HCT is performed on samples with known mass and particle size fraction under unsaturated, aerobic conditions using a weekly cycle of dry air, moist air (e.g., Moyo et al., 2025). In a standardized test such as ASTM D5744 for example, at least 1–2 kg of waste sample (<150 μm particle size for tailings and/or <6.3 mm for waste rock) is exposed to dry air (moisture content, 10%) for 3 days and then to humid air (moisture content of 90%) for another 3 days in a humidity cell. On the 7th day, the waste sample is saturated with some amount of distilled water (water/sample mass is 0.5

or 1.0) for at least 1 h (Koc et al., 2023). Thereafter, leachate is filtered for geochemical analysis. In this way, the acidity (pH, SO_4^{2-}), released alkali ions (Ca, Mn, Mg, Na, Si, Al, K, etc.) and heavy metal ions (Fe, Cu, Zn, Cd, etc.) due to the artificial oxidation of the tailings are analyzed and evaluated (Koc et al., 2023).

By comparing the two laboratory tests, both the HCT and KLC tests are conducted over weeks to years under controlled conditions where water is added to a known quantity of material, and the chemistry of generated leachates is analyzed (e.g., Moyo et al., 2025). According to Koc et al. (2023), the HCT method is advantageous over KLC because (1) less material is used during the experiments and (2) field conditions (temperature, humidity) can be simulated in a more controlled way in the laboratory environment. However, for both the HCT and KLC, the dimensions of the cells or column influence the geochemical response of the analyzed sample (Erguler et al., 2014; Pieretti et al., 2022). For example, Pieretti et al. (2022) performed humidity cell experiments by following the ASTM D5744–18 standard and by using two humidity cells, with distinct shapes and dimensions (Fig. 2.2) packed with the same waste rocks. The results suggest a relatively higher chemical weathering rate (approximately by a factor of 2–3) in the long and narrow shaped humidity cell setup compared to the short and broad cell as indicated by the measured weekly mass loadings for different elements.

2.4.2 Field based tests

Field tests are small-to-large scale trials using, barrels or test piles (e.g., Andrina et al., 2006), for example, Fig. 2.3. Large scale piles require installation of instrumentation including thermistors and lysimeters to monitor the in-pile geochemical conditions (Mertz et al., 2025; Vriens et al., 2020). The benefit of field tests is that mine wastes are exposed to site-relevant climatic and microbiological conditions; therefore field tests are the most accurate method for testing the long-term behavior of mine wastes. However, a significant limitation apart from being costly and time-intensive is that they cannot be established early enough to aid with waste planning at the pre-feasibility or feasibility stages of mine operation. Consequently, the greatest value of field tests is in assessing potential capping and remediation methods of waste storage facilities upon mine closure (e.g., Andrina et al., 2006; Mauric and Lottermoser, 2011).

Figure 2.3 The actual tacked field barrels (left) with a schematic diagram of their internal structure (right). Each field barrel contained three upper lysimeters installed at the contact between two material types, which were designed to sample leachate from the top material before it entered the second material. Drainage from the upper lysimeters all flow into bucket a, whereas drainage that had been in contact with both material types flow into bucket b. The total height of the stacked barrels is 1.8 m. *(From Hirsche et al., 2017).*

2.4.3 Calculations and interpretation of kinetic test results

The leachate collected periodically (mostly in weeks) is subjected to analysis for several parameters, pH, electrical conductivity, acidity alkalinity, sulfide oxidation (acid generating) indicators, and carbonate dissolution (acid neutralizing) indicators. Calcium, Mg, and Mn are indicators of neutralization mineral dissolution while SO_4^{2-} indicates sulfide mineral oxidation (Benzaazoua et al., 2004). Using the amount of aforementioned parameters determined periodically usually weekly and in relation to their initial values in the column/cell, the amount of the analyte that has reacted or weathered can be calculated and consequently their rates of reaction or weathering or depletion be obtained. In many kinetic tests, the volume leached (V) is recorded, and concentration (C) of analyte is determined. With these data and knowledge of reaction taking place (stoichiometry), calculations can be done to identify the sulfide oxidation rate and carbonate dissolution rates. For example, the analyte mass (mg) can be calculated using Eq. 2.32 and can be normalized to their initial weight as per (Eq. 2.33).

$$\begin{aligned}&\text{Released mass per week}\left(\frac{\text{mg}}{\text{week}}\right)\\&=\text{analyte concentration}\left(\frac{\text{mg}}{\text{L}}\right)\times\text{Leachate volume(L)}\end{aligned}\tag{2.32}$$

$$\begin{aligned}&\text{Released rate}(\text{mg/kg/week})\\&=\text{release mass per week}\left(\frac{\text{mg}}{\text{week}}\right)\div\text{sample mass}(\text{kg})\end{aligned}\tag{2.33}$$

Worked example 2

For example, waste rock material was determined to have NP of 48.42 ton $CaCO_3$/1000 ton and total sulfur 0.58%. Given that 1879.2 g of the material was subjected to leaching column to determine the rate of carbonate dissolution and sulfide oxidation rate, calculate the following: (1) mass of sulfur initially present in column and (2) percentage of pyrite initially present in sample.

Solution

Data from static tests are useful in computing the initial amount of the analyte in the column (Eq. 2.34). source: From USEPA, 2011.

(i) initial mass of sulphur in the column (g)

$$= \text{sampleweight (g)} \times \frac{\text{total S(\%)}}{100} \tag{2.34}$$

$$= 1879.2\ \text{g} \times (0.58/100) = 10.90\ \text{g}$$

(ii) pyrite (FeS_2) has a molar mass of 119.97 g/mol, calculated from iron (55.85 g/mol) and two sulfur atoms (32.06 g/mol × 2). Sulfur therefore constitutes 53.45% of pyrite by weight. Accordingly, the pyrite content of a rock can be estimated by multiplying the measured total sulfur (%) by 1.873 (100 ÷ 53.45). Thus the initial pyrite content of the sample is:

$$\%\ \text{pyrite} = \%\ \text{total S} \times 1.873 = 0.58\% \times 1.873 = 1.09\%$$

2.4.3.1 Neutralizing potential depletion time versus acid-producing potential depletion time comparison method

The ASTM method 2013 or HCT procedure compares the time taken for all neutralization materials to react and be deprived in the leachate (NP depletion time) to the time taken for the all-acid generation materials to be oxidized and get deprived in the leachate (AP depletion time). In simple terms, it is the comparison of time taken for NP to be depleted versus the time for AP to be depleted. Calculation of NP depletion time (Eq. 2.35) is based on Ca and Mg generation rates, whereas AP depletion time (Eq. 2.36) is based on SO_4^{2-} production rate (White & Jeffers, 1993). Where the AP and NP are expressed in units of kg$CaCO_3$/t, R_{SO4}, R_{Ca}, and R_{Mg} are release rates of sulfate, calcium, and magnesium, respectively, converted in kg$CaCO_3$/t.

$$\text{Time of NP depletion} = \frac{\text{AP}}{R_{Ca} + R_{mg}} \tag{2.35}$$

$$\text{Time of AP depletion} = \frac{\text{AP}}{R_{SO_4}} \tag{2.36}$$

It follows therefore that if NP is consumed early in the test and AP begins to dominate soon after, the material is considered PAF because its buffering capacity is insufficient to neutralize acidity over time. Conversely, if NP persists throughout the test and AP does not exceed it, the material is classified as NAF (e.g., Plante et al., 2020; USEPA, 2011).

2.4.3.2 Determination of lag time

The time it takes for acidic conditions to develop is called the lag time or time of onset (Erguler et al., 2014; Sapsford et al., 2008; Sexsmith et al., 2015; Stewart, Miller, & Smart, 2006). Lag time is critical for determining when management plans should be implemented to prevent or mitigate ARD/AMD. It is typically determined by theoretical calculations or graphical methods. Calculation method assumes that lag time is the function of depletion of neutralization material and is therefore deduced from Eq. (2.35).

Graphical methods require monitoring pH over time during kinetic tests. A threshold pH is set, and a graph of time (*x*-axis) versus pH (*y*-axis) is plotted; the time it takes to reach this threshold pH represents the lag time. There is no single established threshold pH value for determining lag time. For example, Sexsmith et al. (2015) define actual lag time as the point in time when the pH drops below 5.0. According to Sexsmith et al. (2015), the lag time was selected to ensure that acidic conditions were due to sulfide oxidation rather than equilibrium with atmospheric carbon dioxide. Conversely, Sapsford et al. (2008) use a pH of 4.5 as a boundary to infer whether a sample has the potential to generate net acidity, whereas Erguler et al. (2014) use pH 4.0 as the boundary for acid formation, determining the lag time as the duration until pH drops to 4.0. In their study, Erguler et al. (2014) use lag time to identify parameters affecting the rate of AMD generation during kinetic column tests. They find that the lag time increases with the length, diameter, and volume of the columns for both fine- and coarse-grained particle sizes. For example, the coarse-grained column shows a longer lag time than the fine-grained column (Fig. 2.4).

2.4.3.3 Chemical/mineralogical depletion method

Chemical/mineralogical depletion method is based on calculation of time of depletion of mineral responsible for acid generation (e.g., sulfides) and minerals responsible for neutralization such as carbonates. Depletion time of carbonate minerals is calculated from release rates of Ca, Mg, and Mn (Eq. 2.37), whereas the time for sulfide depletion is calculated from release rate of SO_4^{2-} (Eq. 2.38) where 2.9958 is conversion factor to convert

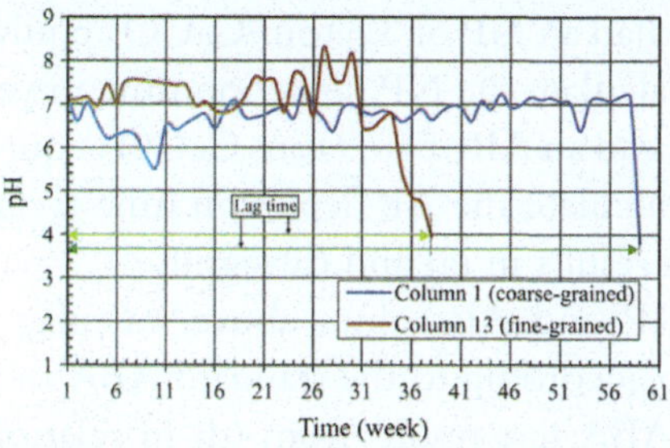

Figure 2.4 Graphical presentation of lag time or the onset of AMD. Lag time defined as the time at which pH drops to 4 showing the effect of grain size on the rate of AMD formation. *AMD*, Acid mine drainage. *(From Erguler et al. (2014)).*

SO_4 to sulfur. $M_{SO_4}$and M_s are molar masses of SO_4^{2-} (g/mol) and atomic masses of sulfur, respectively. R_{SO_4}, R_{Mg}, R_{Ca}, and R_{Mn} (mg/kg/day) are release rates of SO_4^{2-}, Mg, Ca, and Mn, respectively, while $carbonate_{initial}$ and $S_{initial}$ are carbonate and sulfide contents (mg/kg), respectively (Plante et al., 2020).

$$\text{Time for carbonate depletion} = \frac{\text{Carbonate}_{\text{initial}}}{\left(R_{Ca} + R_{Mg} + R_{Mn}\right)} \tag{2.37}$$

$$\text{Time for sulfide depletion} = \frac{M_{SO_4} \times S_{initial}}{M_s R_{SO_4}} = 2.9958 \times S_{initial}/R_{SO_4} \tag{2.38}$$

Under conditions that initial composition of minerals is known precisely, depletion time for each mineral can be calculated using their cations such as Cu, Zn, Ca, and Mg for chalcopyrite, sphalerite, calcite, and dolomite, respectively, while Al and Na can be used for aluminosilicate minerals (Plante et al., 2020). This kind of calculation is called cation approach and is based on three assumptions that (1) all Ca and Mg in solution are derived from carbonate dissolution, (2) Ca and Mg have not been lost from the solution and retained in the column, and (3) gypsum is not present in the material being leached (USEPA, 2011).

Worked example 3

In a study by White and Jeffers (1993), a 1-kg waste rock sample was subjected to accelerated weathering in a humidity cell. The average calcium plus magnesium release rates for weeks 5–21 and 5–51 of leachate collection were 0.30 and 0.25 mill moles per week, respectively. The corresponding sulfate release rates for these periods were 27 and 23 mg/kg/week, respectively.

a If the sample had an NP of 3.3 tons $CaCO_3$ equivalent per 10^3 tons of waste rock, calculate the NP depletion time in years for both periods.
b If the sample had an AP of 46.9 tons $CaCO_3$ equivalent per 10^3 tons of waste rock, calculate the AP depletion time (years), for both periods.
c What do the results in (a) and (b) signify?
d Using the NP and AP values above, calculate the NNP and the NP/AP ratio according to the standard ABA method.
e Discuss the ABA test results from (d) in relation to the kinetic test results from (a) and (b).

Solution

(a) *NP depletion time*

$$NP = \frac{3.3\ \text{tonCaCO}_3}{10^3\text{ton waste rock}} = \frac{3.3\ \text{ton} \times \left(\frac{10^6\ \text{kg}}{\text{ton}}\right) \times \text{CaCO}_3}{\left(10^3\ \text{ton} \times \frac{10^3\ \text{kg}}{\text{ton}}\right)} = \frac{3.3\text{gCaCO}_3}{\text{kg}}$$

$NP=\frac{\left(\frac{3.3\text{g}}{100\text{g/mol}}\right)\text{CaCO}_3}{\text{kg}} = 0.033\ \text{mol}\frac{\text{CaCO}_3}{\text{kg}} = 33\ \text{milmol}\frac{\text{CaCO}_3}{\text{kg}}$; from moles = mass (g) ÷ molar mass (g/mol) where molar mass of $CaCO_3$ is approximately 100 g/mol and 1 mol = 1000 milmol

To calculate NP depletion time in weeks

Recall from (White & Jeffers, 1993), NP depletion time $= \frac{\text{NP(in units of CaCO}_3\text{/t)}}{\text{(release rates of Ca+Mg)}}$

NP depletion time in weeks$=\left(\frac{33\text{milmolCaCO}_3}{\text{kg}}\right)/\left(0.30\frac{\text{milmol}}{\text{week}}\right)=$ 110 weeks; for the weeks 5–21 period

NP depletion time in weeks $= \left(\frac{33\text{milmolCaCO}_3}{\text{kg}}\right)/\left(0.25\frac{\text{milmol}}{\text{week}}\right)=$ 132 weeks; for the (weeks 5–51)

One year is approximately 52 weeks; therefore the NP is 110 weeks $\times \frac{1\ \text{year}}{52\ \text{weeks}} =$ 2.11 years for the weeks 5–21 periond; and 132 weeks $\times \frac{1\ \text{year}}{52\ \text{weeks}} =$ 2.54 years for the weeks 5–51 period

(b) To calculate AP depletion time

Recall from (White & Jeffers, 1993), AP depletion time $= \frac{\text{AP(in units of CaCO}_3\text{/t)}}{\text{(release rates of SO}_4\text{)}}$

$$AP = \frac{46.9\ \text{ton CaCO}_3}{10^3\ \text{ton of wate rock}} = \frac{46.9\ \text{ton} \times (10^9\ \text{mg/ton})}{\left(10^3\ \text{ton} \times 10^3\ \text{kg/ton}\right)} = 46,900\frac{\text{mg}}{\text{kg}}$$

$$\text{AP depletion time} = \frac{\text{AP (in units of CaCO}_3\text{/t)}}{\text{(release rates of SO}_4\text{)}}$$

$$\text{AP depletion time} = \frac{46,900\left(\frac{\text{mg}}{\text{kg}}\right)}{27\ \text{mg/kg/week}}$$

$$\text{AP depletion time} = \frac{46,900\left(\frac{\text{mg}}{\text{kg}}\right)}{23\ \text{mg/kg/week}}$$

Therefore, the AP depletion times are 33.4 years and 39.2 years for the 5–21 and 5–51 week leaching periods, respectively.

(c) *interpretation*

The results from NP and AP depletion times in *a* and *b*, respectively, show that it will take 2–2.5 years for the neutralizing material to be depleted in the studied sample while the material will continue producing acid for 33 years and 39 years based on analyses of humidity cell leachate collected at the periods of 5–21 and 5–51 weeks, respectively. It is portrayed that the material will be acid producing for about three decades after all the neutralizing material has been depleted with the initial 2–2.5 years (White & Jeffers, 1993).

(d) NNP ($kgCaCO_3$) = NP ($kgCaCO_3$) − AP ($kgCaCO_3$)

NNP = NP − AP = 3.3 − 46.9 = −43.6 tons CaCO3/t

NP/AP = 3.3/46.9 = 0.070

(e) The results of ABA test in (d) with the NNP being negative and NP/AP being smaller than 1 show that the material had insufficient acid neutralizing material to neutralize the acid produced sulfide minerals oxidation and therefore the material is classified as acid producing. Kinetic test results in (b) show that rate of acid production will be higher after about 2.5 years when all NP material have completely reacted, whereas acid production will continue until 39 years when all the sulfide will be fully oxidized and no further acid generation. Overall kinetic test shows the risk that this material will cause to the environment if it is left untreated.

Worked Example 4

Using the data in Table 2.3, calculate % $CaCO_3$ weathered each week from an initial sample mass of 91.0 g.

Solution

The solution is reproduced from USEPA, 2011. The percentage of $CaCO_3$ weathered each week, also referred to as the rate of carbonate dissolution, is calculated from the sum of cumulative Ca and Mg expressed

Table 2.3 Weekly leachate volume (mL) and dissolved Ca and Mg concentrations (mg/L) released during the column leaching experiment.

Week	Vol Out mL	mg/L Ca	mg/L mg
0	1356	99.8	57.1
1	310	270.0	148.0
2	340	240.0	131.0
3	295	186.0	93.3
4	309	175.0	82.7
5	270	170.0	78.8
6	279	176.0	83.7
7	296	147.0	68.7
8	285	153.0	68.4
9	285	163.0	84.3
10	268	156.0	68.6
11	260	142.0	62.7
12	260	148.0	59.8
13	274	162.0	68.7
14	264	151.0	66.4

Source: Data from USEPA, 2011.

as $CaCO_3$ equivalents relative to the initial sample weight. Summing the cumulative Ca- and Mg-derived $CaCO_3$ equivalents ensures that all dissolved carbonate material is accounted for, irrespective of whether it originated from calcite or dolomite, assuming that all reactive carbonate occur as Ca and Mg (Table 2.4).

2.4.3.4 Oxidation neutralization curves

Oxidation neutralization curves are expressed as plots of cumulative loads of Ca, Mg, and Mn (neutralization mineral dissolution) leached out of kinetic tests against that of sulfate loads from sulfate oxidation (Benzaazoua et al., 2004). The first part of these curves is graphing sulfate (in mg/kg/week) in x-axis against neutralization products (Ca, Mg and Mn) in y-axis assuming linear relationship in accordance to Eq. (2.39).

Linea regression equation:

$$Y = mx + C \tag{2.39}$$

Y is a predicted value of neutralization for given values of oxidation (x), C is intercept and m = slope (regression coefficient). Several observations can be made from this graph (1) if linear graph is obtained; then, dissolution of carbonate mineral is direct proportional to sulfide oxidation—thus every

Table 2.4 Results for calculation of % $CaCO_3$ weathered each week from an initial sample mass of 91.0 g.

Week	Vol Out mL	mg/L Ca	mg Ca	Cumulative mg Ca	Cumulative mg Ca as $CaCO_3$	mg/L Mg	mg Mg	Cumulative mg Mg	Cumulative mg Mg as $CaCO_3$	Cumulative Ca + Mg as $CaCO_3$	% $CaCO_3$ weathered each week from 91.0 g
0	1356	99.8	135.33	135.33	338.32	57.1	91.83	91.8	377.90	716.22	0.79
1	310	270.0	83.70	219.03	547.57	148.0	54.41	146.24	601.82	1149.39	1.26
2	340	240.0	81.60	300.63	751.57	131.0	52.82	199.07	819.20	1570.77	1.73
3	295	186.0	54.87	355.50	888.75	93.3	32.64	231.71	953.53	1842.28	2.02
4	309	175.0	54.08	409.57	1023.93	82.7	30.31	262.02	1078.25	2102.19	2.31
5	270	170.0	45.90	455.47	1138.68	78.8	25.23	287.25	1182.09	2320.78	2.55
6	279	176.0	49.10	504.58	1261.44	83.7	27.70	314.95	1296.07	2557.51	2.81
7	296	147.0	43.51	548.09	1370.22	68.7	24.12	339.06	1395.32	2765.54	3.04
8	285	153.0	43.61	591.69	1479.24	68.4	23.12	362.18	1490.46	2969.70	3.26
9	285	163.0	46.46	638.15	1595.37	84.3	28.49	390.68	1607.72	3203.09	3.52
10	268	156.0	41.81	679.96	1699.89	68.6	21.80	412.48	1697.45	3397.34	3.73
11	260	142.0	36.92	716.88	1792.19	62.7	19.33	431.82	1777.01	3569.21	3.92
12	260	148.0	38.48	755.36	1888.39	59.8	18.44	450.26	1852.90	3741.29	4.11
13	274	162.0	44.39	799.75	1999.36	68.7	22.33	472.58	1944.77	3944.13	4.33
14	264	151.0	39.86	839.61	2099.02	66.4	20.79	493.37	2030.32	4129.35	4.54

Source: From USEPA, 2011.

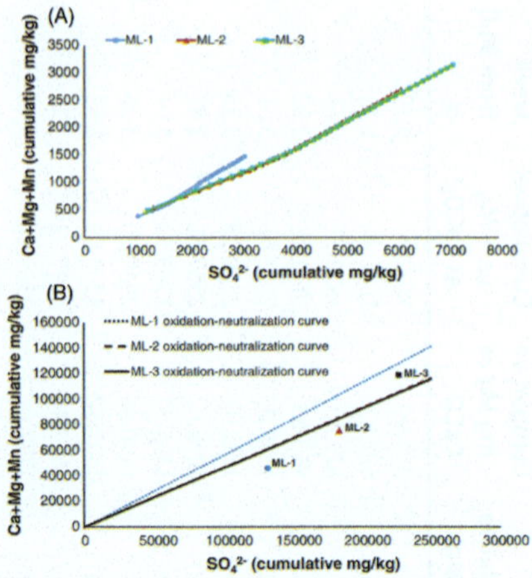

Figure 2.5 ***Prediction of long-term acid generating potential by extrapolation oxidation–neutralization curve*** (A) Cumulative Ca + Mg + Mn versus cumulative sulfate concentration. (B) Extrapolated oxidation–neutralization curve for the acid generating potential long-term prediction *(From Bouzahzah et al. (2014))*.

x mole acid produced is neutralized by *y* mole of carbonate mineral (s) dissolved; (2) the slope of the curve, also called *neutralization rate*, represents the efficiency of the buffering material in neutralizing the acid produced by sulfide oxidation, whereby a steep slope (high Ca + Mg/S ratio) indicating rapid neutralization by carbonates or other neutralizing minerals, corresponding to lower potential for acid generating. When the slope (low Ca + Mg/S ratio) is shallower, it indicates a limited neutralization, suggesting that acidic products may exceed the neutralizing capacity of the rock, which subsequently imply potentially acid generating (Pondja Jr et al., 2017).

Second and most important part of interpretation is long-term prediction of AMD potential by extrapolating the oxidation–neutralization curve and by projecting the initial sulfate (from S content) and Ca, Mg, and Mn concentration in the tailings (Hakkou et al., 2009). If the point falls on the sulfate side of the relationship, then the neutralization mineral will be depleted first suggesting that the material is PAF. In contrast, if point falls on the Ca + Mg + Mn side of the relationship, then the material is nonacid forming (Benzaazoua et al., 2004). For instance, the plot of cumulative Ca + Mg + Mn versus sulfates in Fig. 2.5A is linear and can be extrapolated. Extrapolation of graph line from Fig. 2.5B and locations of samples' the initial solid content show that the samples ML-1 and ML-2 are located below the oxidation–neutralization curves (more toward sulfate axis) are therefore classified as PAF. Sample ML-3 is above its oxidation–neutralization curve and is therefore not acid forming (Bouzahzah et al., 2014).

The hypotheses underlying this method of interpretation are that (1) the geochemical environment stays the same, (2) the ratio between neutralization and oxidation products stays linear, and (3) mineralogy will not affect the linearity of the relation over long time—for instance by depletion or precipitation of mineral (Hakkou et al., 2009). These hypotheses are interdependent and can be realistic as long as conditions of pH, Eh, and T remain the same (Hakkou et al., 2009).

2.4.4 Scaling up of laboratory-based kinetic tests

Scaling up can be defined as transitioning a validated laboratory process to a large-scale operational system while maintaining its efficiency, cost-effectiveness, and consistency at industrial scale. In the context of AMD/ARD, there is no single standardized equation for scaling up laboratory kinetic tests to field scale. Instead, various factors influencing the variability between laboratory tests and between laboratory test and field have been addressed (Elghali et al., 2023; Pieretti et al., 2022).

Elghali et al. (2023) reviewed a total of 12 papers from 2001 to 2022 which discussed the factors that cause scale effect (variability) in kinetic tests. The authors grouped these factors broadly into two main categories: (1) intrinsic factors related to the characteristics of the tested samples such as particle size distribution and mineral liberation degrees and (2) extrinsic factors related to the applied tests such as climatic conditions in laboratory versus field tests, test dimensions (weathering cells vs. humidity cells vs. column tests vs. field tests), liquid–solid ratio, flushing frequency, test duration, water flow (preferential water flow), biological activity, and the required sample mass for each kinetic test. Elghali et al. (2023) found that the scale factor among laboratory tests or between laboratory and field tests is variable and study specific. In addition, the effect of aeration on weathering rate of sulfide and the effect of secondary minerals on scale up of kinetic tests are not known. Moreover, studies have shown that kinetic tests such as humidity cells and leaching columns give inconclusive results for prediction of neutral mine drainage (Demers et al., 2013; Plante et al., 2010; Marmier et al., 2025a).

2.5 Predicting neutral mine drainage

Neutral mine drainage (NMD) is more difficult to predict than AMD because, during pre-mine assessments, contaminants in waste rock or overburden may appear stable (immobilized) and unlikely to leach but can leach out at higher concentrations when the processes that immobilize them become saturated or depleted (Marmier et al., 2025a, 2025b, 2025c). These immobilization processes include sorption of metals or metalloids on mineral surfaces and the precipitation or co-precipitation of metals/metalloids at circumneutral pH (Marmier et al., 2025a, 2025b; Plante et al., 2011a, 2011b, 2012). Therefore, as of 2025, research continues on the geochemistry, likelihood of occurrence, and prevention or treatment methods for NMD (Barago et al., 2023; Pakostova et al., 2023; Marmier et al., 2025c).

Plante et al. (2015) developed a modified kinetic test using a chelating agent instead of water to assess the geochemical behavior of mine wastes at neutral pH. They tested fresh and weathered waste rocks from a Canadian hemo-ilmenite mine to examine the long-term metal release and the impact of ethylenediaminetetraacetic acid (EDTA) on secondary minerals. EDTA is a powerful chelating agent and hexadentate ligand widely used in chemistry to bind metal ions (Fe^{2+}/Fe^{3+}, Ca^{2+}, Mg^{2+}). It forms stable, water-soluble complexes, making it essential for water softening, complexometric titration (hardness testing), cleaning agents, and stabilizing food or cosmetics. The experimental setup included control test where leaching was done by using deionized water, and EDTA-modified tests performed on both fresh and weathered samples, using a 0.03-M EDTA solution in place of water. The results revealed elevated concentrations of leached elements in the presence of EDTA than in deionized water alone demonstrating EDTA's ability to chelate metals immediately upon their release from sulfide oxidation or mineral dissolution. Plante et al. (2015) therefore concluded that their novel method can be used to identify hidden contaminants that would not appear in classic kinetic testing but that would emerge after many years in the field.

Ten years later, Marmier et al. (2025a) modified the Plante et al.'s (2015) method by incorporating a risk assessment on the sorption potential of mining wastes. Their approach assumes that if the mining material has a higher adsorption capacity for a toxic metal/metalloid that exceeds the total concentration of that metal initially present, the mining waste can adsorb the element completely, resulting in a low risk of its drainage at circumneutral pH; conversely, if the adsorption capacity is lower, the risk is higher. The Marmier et al. (2025a) method consists of a combination of three steps to assess the risk of NMD: (1) initial characterization of the mine waste (comprising mineralogical assessment, base, and trace element content and ABA testing); (2) modified column leaching experiments, in which chelating agents are used to determine leaching rates without immobilization processes, in comparison to control column tests using deionized water, in which case data are used to confirm that the mine waste is nonacid generating; and (3) sorption experiments to determine maximum sorption capacity of the mine waste for a given element, assuming sorption is the only mechanism through which contaminants are immobilized before the onset of NMD. This method was applied to study the leaching behavior of metals, for example, Ni and Zn and metalloids, for example, As and Sb, and the findings are presented in Table 2.5.

Table 2.5 shows that the approach of Marmier et al. (2025a) can successfully identify hidden contaminants in NMD. This is revealed by the relatively higher levels of metals and metaloids after using EDTA as a leaching agent to replace deionized water. For instance, concentration of Ni was initially 37 m/L but after leaching the materials with EDTA, it dropped to 9.5 m/L. Similarly, concentration of As ranged from 811m/L to 20.6 in EDTA cells and 1.61 to 0.86 m/L in control (deionized water) cells (Marmier et al., 2025c), which is a similar trend to that of the method of Plante et al. (2015) (Table 2.5). Additionally, the waste rocks can be classified as high risk/low risk of NMD based on the ratio of the waste rock's sorption capacity (q_{max}) to its target element total concentration ($C_{initial}$). For instance, waste rocks from the Lac Tio Mine, Canada, was initially classified as uncertain in the ABA test with $q_{max}/C_{initial}$ ratio of 0.33 reclassified as high risk of leaching Ni from NMD (Marmier et al., 2025a). Another example comes from the gold mine tailings from Eleonore mine, Canada, whereby initial classification indicated this mine as uncertain or not potentially acid generating (NPAG) based on the ABA test but was later on reclassified as high-risk NMD for As following the $q_{max}/C_{initial}$ ratio of 0.08 (Marmier et al., 2025c). In contrast, the four lithologies from Canadian Malaritic mine were classified as NPAG (CPO, AUM, TCH) and uncertain (CGR) but were all later on reclassified as low risk of Zn NMD based on $q_{max}/C_{initial}$ ratios well above 1 (i.e., 17.85 22.9 for CPO, 17.85 for CGR, 10.81 for AUM, and 14.93 for TCH) (Marmier et al., 2025c).

Unlike Marmier et al. (2025a, 2025b), who focused on metal cations, Marmier et al. (2025c) focused on As and Sb oxyanions within the tailings. They found that EDTA can effectively assess NMD of metalloids, where sorption is the primary controlling mechanism for the temporary immobilization of As. Marmier et al. (2025c) found that there is an increase in sorption capacity of the tailings with duration of column testing 43 mg/kg initial sorption capacity to 50 and 76 mg/kg at the end of column testing, suggesting that the tailings continuously generated Fe-oxyhydroxides to (co)

Table 2.5 Assessment of neutral mine drainage in waste rock and tailings using EDTA-modified kinetic leaching method.

	Acid–base accounting (ABA)			NMD source	Target metal	Leaching (mg/L)		Metal content vs. sorption			
Study samples	NP–AP kg$CaCO_3$/t	NP/AP	Classification			With Water	With C_{EDTA}	$C_{initial}$ (mg/kg)	q_{max} (mg/kg)	$q_{max}/C_{initial}$	NMD risk
Waste rocks from Lac Tio Mine, Canada (Marmier et al., 2025a)	2.3	1.4	Uncertain	Ni- and Co-rich pyrite	Ni	0.22–0.007	37–9.5	590	197	0.33	High
Carbonated porphyry (CPO) from Canadian Malaritic mine (Marmier et al., 2025b)	30.70	5.47	NPAG	Sphalerite	Zn	NR	1.00–0.2	54.50	1248	22.90	Low
Carbonated greywacke (CGR) from Canadian Malaritic mine (Marmier et al., 2025b)	13.20	1.89	Uncertain	NR	Zn	NR	1.00–0.2	68.00	1214	17.85	Low

(continued on next page)

Altered ultramafic (AUM) from Canadian Malaritic mine (Marmier et al., 2025b)	69.30	6.89	NPAG	Zn-rich chromite	Zn	NR	1.00–0.2	97.50	1054	10.81	Low
Chlorite schist (TCH) from Canadian Malaritic mine (Marmier et al., 2025b)	106	17.00	NPAG	NR	Zn	NR	NR	68.50	1023	14.93	Low
Gold mine tailings from Eleonore mine, Canada (Marmier et al., 2025c)	6.90	3.10	NPAG/ uncertain	Löllingite and arsenopyrite	As	1.61–0.86	811–20.6	527	43	0.08	High

AP, Acid-producing potential; *EDTA*, ethylenediaminetetraacetic acid; *NMD*, neutral mine drainage; *NP*, neutralizing potential; *NR*=, not reported.

precipitate As (and Sb) which may decrease the intensity of contamination over long term.

Despite the EDTA's method strength in its ability to identify hidden neutral mine drainage risks by bypassing immediate mineral immobilization, it has some limitations (Dieter, 1990; Jez & Lestan, 2016; Sillanpää & Oikari, 1996) which are as follows: (1) it overestimates metal mobilities since EDTA is a strong chelating agent that can cause dissolution of stable secondary minerals (e.g., oxyhydroxides) that not typically occur under natural conditions; (2) EDTA can slightly affect the sulfide oxidation and carbonate dissolution rates, hence affecting natural buffering reactions; and (3) EDTA is a relatively toxic and persistent chemical, which discourages its use, especially in field-scale experiments.

2.6 Predictive models for AMD

One of the models used in AMD prediction is reactive-transport model (RTM). Reactive-transport models are based on mass and energy conservation relations and process-based equations that describe transport (hydraulic advection and dispersion, diffusion, convection) as well as chemical reactions across solids, liquids, and gases (Goddéris, Schott & Brantley, 2019). The RTM simulates the complex interplay of water flow (hydrology), oxygen transport, and chemical reactions (e.g., pyrite oxidation) within mine waste piles used for predicting the long-term or century-long generation of acid (El-Affani et al., 2025; Mertz et al., 2025).

The strengths of RTM for AMD prediction area are as follows: (1) the ability to calculate probable mineral dissolution, precipitation of secondary minerals and sorption, which cannot be well described in conventional kinetic tests (Mayer, Frind & Blowes, 2002; Mertz et al., 2021); (2) the possibility of making long-term projections using short-term calibrated data (Demers et al., 2013; Pabst et al., 2017); and (3) capacity to the simulate complex coupling of mass transport processes (advection, diffusion, dispersion) with chemical reactions (redox reactions) and microbial activities (Mertz et al., 2021, 2025). Because of those strengths, the RTM has become an engineering risk assessment tool offering various application in AMD cases according to Plante et al. (2020), including (1) to better understand geochemical behaviors of mine wastes based on laboratory kinetic tests, and pilot or field test results; (2) to evaluate performance of reclaimed mine sites; (3) to study development of neutral mine drainage; (4) to assess the quality of material for use in reclamation of mine wastes; and (5) to probate mobility of acidic contaminants in aquifers.

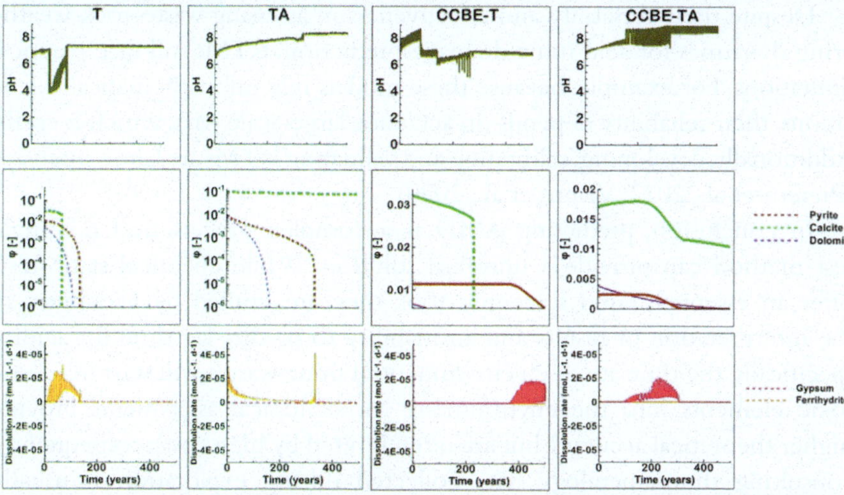

Figure 2.6 ***Application of reactive transport models in AMD prediction.*** A 500 years simulated geochemical evolution of the leachate of low-reactive tailings (T), amended tailings (TA), CCBE-T which is covers with capillary barrier effect (CBBE) whose moisture retaining layer (MRL) is low reactive tailings (without lime amendment); and CCBE-TA is the CCBE with lime amended tailings as MRL. *(From El-Affani et al. (2025).).*

There are several researches employing RTM for predicting AMD. For example, El-Affani et al. (2025) applied a RTM to evaluate the long-term acid-generating potential of low-reactive tailings and the effect of using lime-amended tailings as moisture retaining layer (MRL) in a cover with capillary barrier effects (CCBE) system. The model predicted that tailings could still generate acidity over 70–150 years despite its low sulfur content (2% pyrite), with pH declining to ~4 due to faster calcite depletion relative to pyrite oxidation. In contrast, lime-amended tailings maintained sustained buffering and near-neutral pH; when incorporated into a CCBE system, pH remained above 7 with calcite persisting even after pyrite depletion (Fig. 2.6). Similarly, Mertz et al. (2025) demonstrated that acidity and Pb mobility in altered Pb–Zn tailings are governed by the dissolution and precipitation of secondary minerals, especially anglesite ($PbSO_4$), plumbojarosite ($PbO_5Fe_3(SO_4)_2(OH)_6$), and ferrihydrite. Plumbojarosite dissolution continues for up to 93 years releasing Pb and Fe while promoting anglesite and ferrihydrite precipitation and generates protons that sustain pH around 4. Once all of the plumbojarosite is dissolved the anglesite dissolution starts which also releases Pb according to Mertz et al. (2025).

Despite their versatility and effectiveness in assessing waste-rock weathering dynamics for acid mine drainage prediction, RTMs still face practical limitations. For example, because these models rely on mathematical calculations, their reliability depends on accurate, large-scale data, which is costly. Additionally, developing, calibrating and validating RTMs are labor-intensive (Pieretti et al., 2022; Vriens et al., 2020).

In conclusion, predicting AMD is a complex process, and no single test method can provide a universal solution. While chemical static tests offer an essential, rapid screening tool, they are limited by variability in the interpretation of results and an inability to predict lag time for acidity, specifically the time at which circumneutral mine wastes will start to release toxic elements into the environment. Mineralogical assessments provide higher theoretical accuracy but are often limited by high costs; consequently, consulting the mineralogy data collected during exploration is usually advised. Furthermore, while kinetic tests offer a dynamic view of mine waste behavior, they face challenges in scaling results to field conditions and environmental practicality. Because of the limitations in predicting neutral mine drainage and long-term lag times, field-scale monitoring must remain a priority throughout the operational to postclosure phases to ensure environmental protection.

References

Ackerman, L. áš, Rohovec, J., & Šebek, O. ř (2012). Determination of total sulfur in fifteen geological materials using inductively coupled plasma-optical emission spectrometry (ICP-OES) and combustion/infrared spectrometry. *Geostandards and Geoanalytical Research, 36*(4), 407–414. doi:10.1111/j.1751-908x.2012.00144.x.

Adam, K., Kourtis, A., Gazea, B., & Kontopoulos, A. (1997). Evaluation of static tests used to predict the potential for acid drainage generation at sulphide mine sites. *Transactions of the Institution of Mining and Metallurgy, Section A: Mining Industry, January 1997.*

Akinwekomi, V., Maree, J. P., Masindi, V., Zvinowanda, C., Osman, M. S., Foteinis, S., Mpenyana-Monyatsi, L., & Chatzisymeon, E. (2020). Beneficiation of acid mine drainage (AMD): A viable option for the synthesis of goethite, hematite, magnetite, and gypsum – Gearing towards a circular economy concept. *Minerals Engineering, 148*, 106204. doi:10.1016/j.mineng.2020.106204.

Andersen, C. B. (2002). Understanding carbonate equilibria by measuring alkalinity in experimental and natural systems. *Journal of Geoscience Education, 50*(4), 389–403. Available from http://nagt-jge.org/. doi:10.5408/1089-9995-50.4.389.

Andrina, J., Wilson, G. W., Miller, S., & Neale, A. (2006). Performance of the acid rock drainage mitigation waste rock trial dump at grasberg mine. In *The 7th International Conference on Acid Rock Drainage (ICARD)* (pp. 30–44). https://doi.org/10.21000/JASMR06020030.

Azevedo, R. N. A., Nascimento, A. P., Pereira, V. B., Franco, D. M. M., Nelson, R. K., Gontijo, B., Azevedo, D. A., do Nascimento, R. F., Reddy, C. M., Cavalcante, R. M., Oliveira, A. H. B., & Martins, L. L. (2025). A bottled high-sulfur fuel oil stranded with the mysterious oil spill on the brazilian coast in early 2022: Geochemical correlation and ocean dumping evidence. *Energy and Fuels, 39*(29), 14254–14265. Available from http://pubs.acs.org/journal/enfuem. doi:10.1021/acs.energyfuels.5c01581.

Baek, I., Kim, J., Song, Y., & Kim, T. (2021). Neutralization effect of slag on the acid rock drainage. *International Journal of Geo-Engineering, 12*(1). doi:10.1186/s40703-020-00131-2.

Baker, R. O., Yarranton, H. W., & Jensen, J. L. (2015). Conventional Core Analysis–Rock Properties. *Practical Reservoir Engineering and Characterization* (pp. 197–237). Elsevier. https://doi.org/10.1016/B978-0-12-801811-8.00007-9.

Barago, N. ò, Pavoni, E., Floreani, F., Crosera, M., Adami, G., Lenaz, D., & Covelli, S. (2023). Hydrogeochemistry of thallium and other potentially toxic elements in neutral mine drainage at the decommissioned Pb-Zn Raibl mine (Eastern Alps, Italy). *Journal of Geochemical Exploration, 245*, 107129. doi:10.1016/j.gexplo.2022.107129.

Belzile, N., Chen, Y. W., Cai, M. F., & Li, Y. (2004). A review on pyrrhotite oxidation. *Journal of Geochemical Exploration, 84*(2), 65–76. https://doi.org/10.1016/j.gexplo.2004.03.003.

Benzaazoua, M., Bussière, B., Dagenais, A.-M., & Archambault, M. (2004). Kinetic tests comparison and interpretation for prediction of the Joutel tailings acid generation potential. *Environmental Geology, 46*(8), 1086–1101. https://doi.org/10.1007/s00254-004-1113-1.

Bisutti, I., Hilke, I., & Raessler, M. (2004). Determination of total organic carbon – An overview of current methods. *TrAC Trends in Analytical Chemistry, 23*(10–11), 716–726. doi:10.1016/j.trac.2004.09.003.

Black, S., Allen, D., North, M., Price, B., Rothnie, N., & Sharma, R. (2018). Applicability of the chromium reducible sulfur test for acid metalliferous drainage prediction in hard rock mining. *Applied Geochemistry, 91*, 45–53. doi:10.1016/j.apgeochem.2018.01.013.

Blannin, R., Frenzel, M., Tolosana-Delgado, R., & Gutzmer, J. (2022). Towards a sampling protocol for the resource assessment of critical raw materials in tailings storage facilities. *Journal of Geochemical Exploration, 236*, 106974. doi:10.1016/j.gexplo.2022.106974.

Botero, Y.L., Demers, I., Cisternas, L.A., Ávila, A., & Benzaazoua, M. (2024). A cleaner production strategy for acid mine drainage prevention of waste rock: A porphyry copper case. *International Journal of Mining Science and Technology, 34*(8), 1163–1177. doi:https://doi.org/10.1016/j.ijmst.2024.07.012.

Bouzahzah, H., Benzaazoua, M., & Bussiere, B. (2013). Acid-generating potential calculation using mineralogical static test: Modification of Paktunc equation. *Mise En Garde, 65*.

Bouzahzah H., Benzaazoua M., Bussiere B., Plante B., Prediction of Acid Mine Drainage: Importance of mineralogy and the test protocols for static and kinetic tests . Mine Water and the Environment. **33** (1), (2014), 54–65, doi: 10.1007/s10230-013-0249-1.

Bouzahzah, H., Benzaazoua, M., Plante, B., & Bussiere, B. (2015). A quantitative approach for the estimation of the " fi zz rating " parameter in the acid-base accounting tests : A new adaptations of the Sobek test. *Journal of Geochemical Exploration, 153*, 53–65. https://doi.org/10.1016/j.gexplo.2015.03.003.

Bouzahzah, H., Benzaazoua, M., & Bussiere, B. (2013). Acid-generating potential calculation using mineralogical static test: Modification of Paktunc equation. *Mise En Garde, 65*.

Bouzahzah, H., Benzaazoua, M., Bussiere, B., & Plante, B. (2014). *Prediction of Acid Mine Drainage: Importance of Mineralogy and the Test Protocols for Static and Kinetic Tests. Mine Water and the Environment, 33*(1), 54–65. https://doi.org/10.1007/s10230-013-0249-1.

Bozkaya, O. gün, & Aluç, Y. aş (2024). Physico-chemical characterization of food grade natural spring salt from the Central Anatolia region of Turkey and investigation of its microplastic content. *Journal of Food Science and Technology, 61*(9), 1711–1721. doi:10.1007/s13197-024-05942-0.

Burton, E. D., Sullivan, L. A., Bush, R. T., Johnston, S. G., & Keene, A. F. (2008). *Applied Geochemistry A simple and inexpensive chromium-reducible sulfur method for acid-sulfate soils, 23*, 2759–2766. https://doi.org/10.1016/j.apgeochem.2008.07.007.

Casagrande, D. J., Finkelman, R. B., & Caruccio, F. T. (1989). The non-participation of organic sulfur in acid mine drainage generation. *Environmental Geochemistry and Health, 11*(3-4), 187–192. doi:10.1007/BF01758669.

Chai, Y., Qin, P., Zhang, J., Wu, Z., Li, T., Xu, W., & Sun, H. (2020). Experimental study and application of dolomite aeration oxidation filter bed for the treatment of acid mine drainage. *Minerals Engineering, 157*, 106560. doi:10.1016/j.mineng.2020.106560.

Chopard, A., Benzaazoua, M., Plante, B., Bouzahzah, H., & Marion, P. (2015). Kinetic tests to evaluate the relative oxidation rates of various sulfides and sulfosalts. In *Proceedings of the 10th International Conference on Acid Rock Drainage (ICARD) & IMWA Annual Conference*. Santiago, Chile *April*, 10.

Chopard, A., Benzaazoua, M., Plante, B., Bouzahzah, H., & Marion, P. (2015). Kinetic tests to evaluate the relative oxidation rates of various sulfides and sulfosalts. *Proceedings of the 10th International Conference on Acid Rock Drainage (ICARD) & IMWA Annual Conference, Santiago, Chile, April*, 10.

Claff, S. R., Sullivan, L. A., Burton, E. D., & Bush, R. T. (2010). A sequential extraction procedure for acid sulfate soils: Partitioning of iron. *Geoderma, 155*(3–4), 224–230. https://doi.org/10.1016/j.geoderma.2009.12.002.

Coller, M. E., & Lelnlnger, R. K. (1955). Determination of Total Sulfur Content of Sedimentary Rocks by a Combustion Method. *Analytical Chemistry, 27*(6), 949–954. https://doi.org/10.1021/ac60102a022.

Cornwell, J. C., & Morse, J. W. (1987). The characterization of iron sulfide minerals in anoxic marine sediments. *Marine Chemistry, 22*(2–4), 193–206. https://doi.org/10.1016/0304-4203(87)90008-9.

Czerewko, M. A., Cripps, J. C., Reid, J. M., & Duffell, C. G. (2003). Sulfur species in geological materials – Sources and quantification. *Cement and Concrete Composites, 25*(7), 657–671. doi:10.1016/S0958-9465(02)00066-5.

Davani, B., Russo, K. A., Wilk, A., & Bhattacharyya, L. (2005). Common pharmacopeial calculations in USP monographs. *Pharmacopeial Forum, 31*(2), 626–636.

Davis, A., Whitehead, C., Lengke, M., & Collord, J. (2019). Acid–base accounting tests in combination with humidity cells help to predict waste rock behavior. *Mine Water and the Environment., 38*(3), 467–487. doi:10.1007/s10230-019-00617-1.

Demers, I., Molson, J., Bussière, B., & Laflamme, D. (2013). Numerical modeling of contaminated neutral drainage from a waste-rock field test cell. *Applied Geochemistry, 33*, 346–356. doi:10.1016/j.apgeochem.2013.02.025.

Dhir, B. (2018). *Biotechnological tools for remediation of acid mine drainage (removal of metals from wastewater and leachate) bio-geotechnologies for mine site rehabilitation* (pp. 67–82). Elsevier Available from.http://www.sciencedirect.com/book/9780128129869. doi:10.1016/B978-0-12-812986-9.00004-X.

Dieter, H. H. (1990). Implications of heavy metal toxicity related to EDTA exposure. *Toxicological & Environmental Chemistry, 27*(1-3), 91–95. doi:10.1080/02772249009357561.

Downing, B. W. (2014). Acid Drainage Sampling Design and Sample Preparation. *Acid Mine Drainage, Rock Drainage, and Acid Sulfate Soils* (9780470487, pp. 123–130). Wiley. https://doi.org/10.1002/9781118749197.ch11.

El-Affani, L., Bussière, B., Qureshi, A., & Plante, B. (2025). Geochemical behavior of amended and non-amended mine tailings as cover materials for acid mine drainage control: Column tests and reactive transport modeling. *Journal of Contaminant Hydrology*, (March), 272. https://doi.org/10.1016/j.jconhyd.2025.104564.

Elghali, A., Benzaazoua, M., Taha, Y., Amar, H., Ait-khouia, Y., Bouzahzah, H., & Hakkou, R. (2023). Prediction of acid mine drainage: Where we are. *Earth-Science Reviews, 241*, 104421. doi:10.1016/j.earscirev.2023.104421.

Engleman, E. E., Jackson, L. L., & Norton, D. R. (1985). Determination of carbonate carbon in geological materials by coulometric titration. *Chemical Geology, 53*(1-2), 125–128. doi:10.1016/0009-2541(85)90025-7.

Erguler, G. K., Erguler, Z. A., Akcakoca, H., & Ucar, A. (2014). The effect of column dimensions and particle size on the results of kinetic column test used for acid mine drainage (AMD) prediction. *Minerals Engineering, 55*, 18–29. https://doi.org/10.1016/j.mineng.2013.09.008.

Erickson, P. M., & Hedin, R. S. (1988). *Evaluation of Overburden Analytical Methods as Means to Predict Post-Mining Coal Mine Drainage Quality*. U.S. Center for Disease Control and Prevention (CDC) https://stacks.cdc.gov/view/cdc/219573.

Evangelou, V. P. (Bill), & Zhang, Y. L. (1995). A review: Pyrite oxidation mechanisms and acid mine drainage prevention. *Critical Reviews in Environmental Science and Technology, 25*(2), 141–199. https://doi.org/10.1080/10643389509388477.

Ezz-Eldin, F. M., Abd-Elaziz, T. D., & Elalaily, N. A. (2010). Effect of dilute HF solutions on chemical, optical, and mechanical properties of soda–lime–silica glass. *Journal of Materials Science, 45*(21), 5937–5949. doi:10.1007/s10853-010-4679-x.

Fabin, M., & Jarosz, T. (2025). Perchlorates as emerging contaminants: Sources, hazards, detection and remediation—An overview. *Chemosphere, 387*, 144533. doi:10.1016/j.chemosphere.2025.144533.

Frostad, S.R., Price, W.A., & Bent, H. (2003). Operational NP determination—accounting for iron manganese carbonates and developing a site-specific fizz rating. *Mining and the Environment, Sudbury, 5*(306), 231–237.

Gao, S., Li, J.-, Zhang, S., & Poon, C. S (2025). A state-of-the-art review of solid waste leaching mechanisms and evaluation methodologies. *Waste Management, 204*, 114941. doi:10.1016/j.wasman.2025.114941.

Gibbs, R. J. (1977). Effect of combustion temperature and time, and of the oxidation agent used in organic carbon and nitrogen analyses of sediments and dissolved organic material. *Journal of Sedimentary Research, 47*(2), 547–550. https://doi.org/10.1306/212F71D5-2B24-11D7-8648000102C1865D.

Giraldo, C. V., Acevedo, S. E., Contreras, C. P., Santibáñez, F., Sáez, E., Calderón, F. J., & Bonilla, C. A. (2024). Effects of soil heating changes on soil hydraulic properties in Central Chile. *Geoderma, 449*. Available from https://www.sciencedirect.com/science/journal/00167061. doi:10.1016/j.geoderma.2024.117013.

Goddéris, Y., Schott, J., & Brantley, S. L. (2019). Reactive transport models of weathering. *Elements, 15*(2), 103–106. Available from https://pubs.geoscienceworld.org/elements/article-pdf/15/2/103/4677865/gselements-15-2-103.pdf. doi:10.2138/gselements.15.2.103.

Gomo, M. (2018). Conceptual hydrogeochemical characteristics of a calcite and dolomite acid mine drainage neutralised circumneutral groundwater system. *Water Science, 32*(2), 355–361. doi:10.1016/j.wsj.2018.05.004.

Goumih, A., El Adnani, M., Hakkou, R., Benzaazoua, M., Ouhamdouch, S., & Boumehdi, M. A. (2022). Evaluation of the long-term contaminated neutral drainage CND generation potential of waste rock piles at the abandoned Zn-Pb Erdouz Mine (Occidental High Atlas, Morocco). *Mining, Metallurgy and Exploration, 39*(2), 643–654. Available from https://link.springer.com/journal/volumesAndIssues/42461. doi:10.1007/s42461-022-00553-9.

Hageman, P. L., Seal, R. R., Diehl, S. F., Piatak, N. M., & Lowers, H. A. (2015). Evaluation of selected static methods used to estimate element mobility, acid-generating and acid-neutralizing potentials associated with geologically diverse mining wastes. *Applied Geochemistry, 57*, 125–139. Available from http://www.journals.elsevier.com/applied-geochemistry. doi:10.1016/j.apgeochem.2014.12.007.

Hageman, P. L. (2005). A Simple Field Leach Test to Assess Potential Leaching of Soluble Constituents from Mine Waste, Soils, and other Geologic Materials. *U.S. Geological Survey Fact Sheet 2005-3100, 2005*. https://doi.org/10.21000/JASMR06022639.

Hakkou, R., Benzaazoua, M., & Bussiere, B. (2009). Laboratory Evaluation of the Use of Alkaline Phosphate Wastes for the Control of Acidic Mine Drainag. *Mine Water and the Environment*. https://doi.org/10.1007/s10230-009-0081-9.

Hirsche, D. T., Blaskovich, R., Mayer, K. U., & Beckie, R. D. (2017). A study of Zn and Mo attenuation by waste-rock mixing in neutral mine drainage using mixed-material field barrels and humidity cells. *Applied Geochemistry, 84*, 114–125. Available from http://www.journals.elsevier.com/applied-geochemistry. doi:10.1016/j.apgeochem.2017.06.005.

Hubbard, W. N., Katz, C., & Waddington, G. (1954). Heats of combustion of some sulfur-containing compounds. *Journal of Physical Chemistry, 58*(2), 142–152. doi:10.1021/j150512a010.

Jamieson, H. E., Walker, S. R., & Parsons, M. B. (2015). Mineralogical characterization of mine waste. *Applied Geochemistry, 57*, 85–105. Available from http://www.journals.elsevier.com/applied-geochemistry. doi:10.1016/j.apgeochem.2014.12.014.

Jennings, S. R., Dollhopf, D. J., & Inskeep, W. P. (2000). Acid production from sulfide minerals using hydrogen peroxide weathering. *Applied Geochemistry, 15*(2), 235–243. Available from http://www.journals.elsevier.com/applied-geochemistry. doi:10.1016/S0883-2927(99)00041-4.

Jez, E., & Lestan, D. (2016). EDTA retention and emissions from remediated soil. *Chemosphere, 151*, 202–209. doi:10.1016/j.chemosphere.2016.02.088.

Kanrar, B., Sanyal, K., & Dhara, S. (2021). Quantification and distribution of trace elements in fusion bead and pressed pellet specimens using a table top micro-X-ray fluorescence spectrometer. *Spectrochimica Acta Part B: Atomic Spectroscopy, 177*, 106063. doi:10.1016/j.sab.2020.106063.

Karaca, S., Akyürek, M., & Bayrakçeken, S. (2003). The removal of pyritic sulfur from Aşkale lignite in aqueous suspension by nitric acid. *Fuel Processing Technology, 80*(1), 1–8. https://doi.org/10.1016/S0378-3820(02)00026-7.

Karlsson, T., Räisänen, M. L., Lehtonen, M., & Alakangas, L. (2018). Comparison of static and mineralogical ARD prediction methods in the Nordic environment. *Environmental Monitoring and Assessment, 190*(12). Available from https://link.springer.com/journal/10661. doi:10.1007/s10661-018-7096-2.

Karlsson, T., Alakangas, L., Kauppila, P., & Räisänen, M. L. (2021). A test of two methods for waste rock drainage quality prediction: Aqua regia extraction and single-addition net-acid generation test leachate analysis. *Mine Water and the Environment, 40*(3), 736–751. Available from http://link.springer.de/link/service/journals/10230/index.htm. doi:10.1007/s10230-021-00784-0.

Karlsson, T., Kauppila, P. M., Lehtonen, M., Alakangas, L., & Kauppila, T. (2022). Sulfur analyses and mineralogical data in the preliminary mine waste characterization. *Environmental Monitoring and Assessment, 194*(6), 437. https://doi.org/10.1007/s10661-022-10094-9.

Kieber, R. J., Peake, B., Willey, J. D., & Avery, G. B. (2002). Dissolved organic carbon and organic acids in coastal New Zealand rainwater. *Atmospheric Environment, 36*(21), 3557–3563. doi:10.1016/S1352-2310(02)00273-X.

Kim, Y. uC., & Yoon, H. (2021). Exploitation of acetic acid for calcite dissolution in small-capacity desalination plants. *Desalination, 516*, 115227. doi:10.1016/j.desal.2021.115227.

Klingshirn, C. D., West, Z. J., DeWitt, M. J., Higgins, A., Graham, J., & Corporan, E. (2019). Quantification of elemental and total carbon in combustion particulate matter using thermal-oxidative analysis. *Journal of the Air and Waste Management Association, 69*(8), 1003–1013. Available from http://www.tandfonline.com/loi/uawm20?. doi:10.1080/10962247.2019.1630025.

Koc, E., Cihangir, F., & Ercikdi, B. (2023). Geochemical evaluation of sulfidic tailings and cemented paste backfill with respect to environmental impacts. In C Qi, & C. H. Benson (Eds.), *Managing Mining and Minerals Processing Wastes* (pp. 47–70). Elsevier. https://doi.org/10.1016/B978-0-323-91283-9.00003-1.

Kwong, J.Y. (1993). Prediction and prevention of acid rock drainage a geological and mineralogical perspective. *MEND Project 1.32.1.*

Laban, K. L., & Atkin, B. P. (2000). The direct determination of the forms of sulphur in coal using microwave digestion aid i.c.p.-a.e.s. analysis. *Fuel, 79*, 173–180. https://doi.org/10.1016/S0140-6701(00)92917-5.

Lapakko, K. A. (1994). Evaluation of neutralization potential determinations for metal mine waste and a proposed alternative. *Journal American Society of Mining and Reclamation, 1994*(1), 129–137. doi:10.21000/jasmr94010129.

Lapakko, K. (2002). Metal mine rock and waste characterization tools: An overview. *Mining, Minerals and Sustainable Development*, (Issue 67). https://www.iied.org/g00559.

Lawrence, R. W., & Scheske, M. (1997). A method to calculate the neutralization potential of mining wastes. *Environmental Geology, 32*(2), 100–106. https://doi.org/10.1007/s002540050198.

Lawrence, R. W., & Wang, Y. (1996). *Determination of neutralization potential for acid rock drainage prediction MEND Project 1.16.3*.

Lawrence, R. W., & Wang, W. (1997). Determination of Neutralization Potential in the Prediction of Acid Rock Drainage. In B In (Ed.), *Proceedings of 4th International Conference on Acid Rock Drainage* (pp. 449–464).

León, R., Macías, F., Cánovas, C. R., Millán-Becerro, R., Romero-Matos, J., & Nieto, J. M. (2024). Origin of rare earth elements in acid mine drainage traced by strontium and neodymium isotopes. *Geochimica et Cosmochimica Acta, 372*, 101–110. Available from https://www.sciencedirect.com/science/journal/00167037. doi:10.1016/j.gca.2024.03.025.

Li, J., Smart, R. S. C., Schumann, R. C., Gerson, A. R., & Levay, G. (2007). A simplified method for estimation of jarosite and acid-forming sulfates in acid mine wastes. *Science of the Total Environment, 373*(1), 391–403. doi:10.1016/j.scitotenv.2006.11.012.

Manchisi, J., & Ndlovu, S. (2021). Prediction of Acid Mine Drainage Formation. In G. S. Simate, & S. Ndlovu (Eds.), *Acid Mine Drainage: From Waste to Resources* (1st) (pp. 31–56). CRC Press.

Marmier, V., Plante, B., Demers, I., & Benzaazoua, M. (2025a). Development of a Neutral Mine Drainage Prediction Method Using Modified Kinetics Tests and Assessment of Sorption Capacities. *Mine Water and the Environment, 44*(1), 16–29. https://doi.org/10.1007/s10230-025-01023-6.

Marmier, V., Plante, B., Demers, I., & Benzaazoua, M. (2025b). Neutral Mine Drainage prediction for different waste rock lithologies – Case study of Canadian Malartic. *Journal of Geochemical Exploration, 271*, 107685. https://doi.org/10.1016/j.gexplo.2025.107685.

Marmier, V., Plante, B., Demers, I., & Benzaazoua, M. (2025c). The use of EDTA leaching method to predict arsenic and antimony Neutral Mine Drainage from the Eleonore tailings. *Journal of Geochemical Exploration, 273*(September 2024), 107734. https://doi.org/10.1016/j.gexplo.2025.107734.

Matsumoto, S., Ishimatsu, H., Shimada, H., Sasaoka, T., & Kusuma, G. J. (2018). Characterization of mine waste and acid mine drainage prediction by simple testing methods in terms of the effects of sulfate-sulfur and carbonate minerals. *Minerals, 8*(9). Available from http://www.mdpi.com/2075-163X/8/9/403/pdf. doi:10.3390/min8090403.

Mauric, A., & Lottermoser, B. G. (2011). Phosphate amendment of metalliferous waste rocks, Century Pb-Zn mine, Australia: Laboratory and field trials. *Applied Geochemistry, 26*(1), 45–56. https://doi.org/10.1016/j.apgeochem.2010.11.002.

Mayer, K. U., Frind, E. O., & Blowes, D. W. (2002). Multicomponent reactive transport modeling in variably saturated porous media using a generalized formulation for kinetically controlled reactions. *Water Resources Research, 38*(9). Available from http://agupubs.onlinelibrary.wiley.com/hub/journal/10.1002/(ISSN)1944-7973/. doi:10.1029/2001wr000862.

McLemore, Virginia, T., Smith, Kathleen, S., Russell, Carol, C., et al. (2007). Sampling and monitoring for closure. *Geological Society of America United States*, 171–180. doi:10.1130/2007.4017(11).

Mertz, S., Le Forestier, L., Bataillard, P., & Devau, N. (2021). Leaching of trace metals (Pb) from contaminated tailings amended with iron oxides and manure: New insight from a modelling approach. *Chemical Geology, 579*. Available from http://www.sciencedirect.com/science/journal/00092541. doi:10.1016/j.chemgeo.2021.120356.

Mertz, S., Devau, N., Thouin, H., Battaglia-Brunet, F., Norini, M.-P., Crampon, M., & Forestier, L. Le (2025). Leaching of pollutant metals (Pb, Zn) from abandoned mine tailings: A multicomponent reactive transport model of a pilot-scale experiment. *Science of the Total Environment, 960*, 178248. doi:10.1016/j.scitotenv.2024.178248.

Morin, K. A., & Hutt, N. M. (1994). Observed Preferential Depletion of Neutralization Potential over Sulfide Minerals in Kinetic Tests: Site-Specific Criteria for Safe NP/AP Ratios. In *Journal American Society of Mining and Reclamation* (Vol. 1994, Issue 1, pp. 148–156). https://doi.org/10.21000/jasmr94010148.

Morse, J. W., & Cornwell, J. C. (1987). Analysis and distribution of iron sulfide minerals in recent anoxic marine sediments. *Marine Chemistry, 22*(1), 55–69. https://doi.org/10.1016/0304-4203(87)90048-X.

Moyo, A., Parbhakar-Fox, A., Meffre, S., & Cooke, D. R. (2025). An accelerated kinetic leach test for geochemical and environmental characterisation of acid and metalliferous drainage. *Environmental Technology & Innovation, 38*, 104092. doi:10.1016/j.eti.2025.104092.

Mugova, E., & Wolkerdorfer, C. (2011). *Insights into the First Flush Effect-The Flooded Underground Mine as a Chemical Reactor, 1999*, 305–310.

Murcia, D. C. F., Fosbøl, P. L., Thomsen, K., & Stenby, E. H. (2017). Determination of Zinc Sulfide Solubility to High Temperatures. *Journal of Solution Chemistry, 46*(9–10), 1805–1817. https://doi.org/10.1007/s10953-017-0648-1.

Narasagoudar, R. A., Johnson, J. W., & O'Keefe, T. J (1982). The anodic dissolution of ZnS electrodes in sulfuric acid solutions. *Hydrometallurgy, 9*(1), 37–55. https://doi.org/10.1016/0304-386X(82)90052-4.

Neupane, G., & Donahoe, R. J. (2013). Leachability of elements in alkaline and acidic coal fly ash samples during batch and column leaching tests. *Fuel, 104*, 758–770. doi:10.1016/j.fuel.2012.06.013.

Oh, C., Ji, S., Yim, G., & Cheong, Y. (2017). Evaluation of net acid generation pH as a single indicator for acid forming potential of rocks using geochemical properties. *Environmental Monitoring and Assessment, 189*(4). doi:10.1007/s10661-017-5869-7.

Olds, W. E., Weber, P. A., Pizey, M. H., & Pope, J. (2016). Acid mine drainage analysis for the Reddale Coal Mine, Reefton, New Zealand. *New Zealand Journal of Geology and Geophysics, 59*(2), 341–351. Available from http://www.tandfonline.com/loi/tnzg20?open=56&repitition=0#vol_56. doi:10.1080/00288306.2016.1148056.

Pabst, T., Molson, J., Aubertin, M., & Bussière, B. (2017). Reactive transport modelling of the hydro-geochemical behaviour of partially oxidized acid-generating mine tailings with a monolayer cover. *Applied Geochemistry, 78*, 219–233. Available from http://www.journals.elsevier.com/applied-geochemistry. doi:10.1016/j.apgeochem.2017.01.003.

Pakostova, E., Hilger, D. M., Blowes, D. W., & Ptacek, C. J. (2023). Microbial processes with the potential to mobilize As from a circumneutral-pH mixture of flotation and roaster tailings. *Scientific Reports, 13*(1). Available from https://www.nature.com/srep/. doi:10.1038/s41598-023-50435-3.

Paktunc, A. D. (1999a). Characterization of mine wastes for prediction of acid mine drainage. *Environmental Impacts of Mining Activities* (pp. 19–40). https://doi.org/10.1007/978-3-642-59891-3_3.

Paktunc, A. D. (1999b). Discussion of "A method to calculate the neutralization potential of mining wastes" by Lawrence and Scheske. *Environmental Geology, 38*(June), 82–84.

Paktunc, A. D. (1999). Mineralogical constraints on the determination of neutralization potential and prediction of acid mine drainage. *Environmental Geology, 39*(2), 103–112. https://doi.org/10.1007/s002540050440.

Paktunc, A.D. (1999c). *Mineralogical constraints on the determination of neutralization potential and prediction of acid. 39*(December), 103–112.

Parbhakar-Fox, A., & Lottermoser, B. G. (2015). A critical review of acid rock drainage prediction methods and practices. *Minerals Engineering, 82*, 107–124. Available from http://www.journals.elsevier.com/minerals-engineering/. doi:10.1016/j.mineng.2015.03.015.

Parbhakar-Fox, A. K., Edraki, M., Walters, S., & Bradshaw, D. (2011). Development of a textural index for the prediction of acid rock drainage. *Minerals Engineering, 24*(12), 1277–1287. https://doi.org/10.1016/j.mineng.2011.04.019.

Pawlak, Z., & Pawlak, A. S. (1999). Modification of iodometric determination of total and reactive sulfide in environmental samples. *Talanta, 48*(2), 347–353. https://doi.org/10.1016/S0039-9140(98)00253-7.

Peterson, M. L., Lang, S. Q., Aufdenkampe, A. K., & Hedges, J. I. (2003). Dissolved organic carbon measurement using a modified high-temperature combustion analyzer. *Marine Chemistry, 81*(1–2), 89–104. Available from www.elsevier.com/inca/publications/store/5/0/3/3/4/9. doi:10.1016/S0304-4203(03)00011-2.

Pieretti, M., Karlsson, T., Arvilommi, S., & Muniruzzaman, M. (2022). Challenges in predicting the reactivity of mine waste rocks based on kinetic testing: Humidity cell tests and reactive transport modeling. *Journal of Geochemical Exploration, 237*, 106996. doi:10.1016/j.gexplo.2022.106996.

Plante, B., Benzaazoua, M., Bussière, B., Biesinger, M. C., & Pratt, A. R. (2010). Study of Ni sorption onto Tio mine waste rock surfaces. *Applied Geochemistry, 25*(12), 1830–1844. doi:10.1016/j.apgeochem.2010.09.010.

Plante, B., Benzaazoua, M., & Bussière, B. (2011a). Kinetic testing and sorption studies by modified weathering cells to characterize the potential to generate contaminated neutral drainage. *Mine Water and the Environment, 30*(1), 22–37. https://doi.org/10.1007/s10230-010-0131-3.

Plante, B., Benzaazoua, M., & Bussière, B. (2011b). Predicting geochemical behaviour of waste rock with low acid generating potential using laboratory kinetic tests. *Mine Water and the Environment, 30*(1), 2–21. https://doi.org/10.1007/s10230-010-0127-z.

Plante, B., Bussière, B., & Benzaazoua, M. (2012). Static tests response on 5 Canadian hard rock mine tailings with low net acid-generating potentials. *Journal of Geochemical Exploration, 114*, 57–69. doi:10.1016/j.gexplo.2011.12.003.

Plante, B., Benzaazoua, M., Bussière, B., Kandji, E. H. B., Chopard, A., & Bouzahzah, H. (2015). Use of EDTA in modified kinetic testing for contaminated drainage prediction from waste rocks: case of the Lac Tio mine. *Environmental Science and Pollution Research, 22*(10), 7882–7896. Available from http://www.springerlink.com/content/0944-1344. doi:10.1007/s11356-015-4106-6.

Plante, B., Schudel, G., & Benzaazoua, M. (2020). Prediction of acid mine drainage. In B. Bussière, & M. Guittonny (Eds.), *Hard Rock Mine Reclamation: From Prediction to Management of Acid Mine Drainage* (1st ed.). CRC Press https://doi.org/10.1201/9781315166698.

Plumlee, G. S., Morman, S. A., Meeker, G. P., Hoefen, T. M., Hageman, P. L., & Wolf, R. E. (2013). The environmental and medical geochemistry of potentially hazardous materials produced by disasters. *Treatise on Geochemistry: Second Edition., 11*, 257–304. Available from http://www.sciencedirect.com/science/book/9780080983004. doi:10.1016/B978-0-08-095975-7.00907-4.

Pondja Jr, E.A., Persson, K.M., & Matsinhe, N.P. (2017). Assessment of coal mine water in Moatize by static and leaching tests. *Sustainable Water Resources Management, 3*(4), 403–412. https://doi.org/10.1007/s40899-017-0106-7.

Price, W. A. (2009). *Prediction manual for drainage chemistry from sulphidic geologic materials. CANMET, Mining and Mineral Sciences Laboratories.*

Robinson, P. (2025). GGR handbook of rock and mineral analysis chapter 2 (part 1) analysis of geological materials – 1: Sample preparation methods. *Geostandards and Geoanalytical Research, 49*(1), 9–50. doi:10.1111/ggr.12555.

Sandan, S., & Nurcholis, M. (2024). Analysis of PAF and NAF Tests Using NAPP and ANC Methods at Hajak Site, North Barito, Central Borneo. *Journal of Earth and Marine Technology (JEMT), 4*(2), 270–277. https://doi.org/10.31284/j.jemt.2024.v4i2.5336.

Sapsford, D., Bowell, R., & Dey, M. (2008). A comparison of kinetic NAG tests with static and humidity cell tests for the prediction of ARD. *Mine Water and the Environment, 2*, 325–328. http://www.imwa.info/docs/imwa_2008/IMWA2008_130_Sapsford.pdf.

Sapsford, D. J., Bowell, R. J., Dey, M., & Williams, K. P. (2009). Humidity cell tests for the prediction of acid rock drainage. *Minerals Engineering, 22*(1), 25–36. doi:10.1016/j.mineng.2008.03.008.

Schumann, R., Stewart, W., Miller, S., Kawashima, N., Li, J., & Smart, R. (2012). Acid–base accounting assessment of mine wastes using the chromium reducible sulfur method. *Science of The Total Environment, 424*, 289–296. doi:10.1016/j.scitotenv.2012.02.010.

Sexsmith, K., Macgregor, D., & Barnes, A. (2015). Comparison of Actual and Calculated Lag Times in Humidity Cell Tests. In *10th International Conference on Acid Rock Drainage & IMWA Annual Conference* (pp. 1–10).

Shamrikova, E., Vanchikova, E., Kyzyurova, E., & Zhangurov, E. (2024). Methods for Measuring Organic Carbon Content in Carbonate-Containing Soils : A Review. *Eurasian Soil Science, 57*, 380–394. https://doi.org/10.1134/S1064229323603104.

Sillanpää, M., & Oikari, A. (1996). Assessing the impact of complexation by EDTA and DTPA on heavy metal toxicity using microtox bioassay. *Chemosphere, 32*(8), 1485–1497. doi:10.1016/0045-6535(96)00057-4.

Skousen, J., Renton, J., Brown, H., Evans, P., Leavitt, B., Brady, K., Cohen, L., & Ziemkiewicz, P. (1997). Neutralization potential of overburden samples containing siderite. *Journal of Environmental Quality, 26*(3), 673–681. doi:10.2134/jeq1997.00472425002600030012x.

Smart, R., Skinner, B., Levay, G., Gerson, A., Thomas, J., Sobieraj, H., Schumann, R., Weisener, C., & Weber, P. (2002). *ARD test handbook: project P387A, prediction and kinetic control of acid mine drainage.* AMIRA International Limited www.amira.com.au.

Sobek, A., Schuller, W. A., Freeman, J. R., & Smith, R. M. (1978). *Field and Laboratory Methods Applicable to Overburdens and Minesoils. U.S. Environmental Protection Agency.*

SRKSteffen, Robertson and Kirsten (B.C.) Inc. (1989). Report Prepared for British Columbia Acid Mine Drainage Task Force. *Bi-Tech Publishers Ltd., Vancouver, BC, I.*

Steinmaus, C., Moore, L., Hopenhayn-Rich, C., Biggs, M. Lou, & Smith, A. H (2000). Arsenic in Drinking Water and Bladder Cancer: Environmental Carcinogenesis. *Cancer Investigation, 18*(2), 174–182. https://doi.org/10.3109/07357900009038249.

Stewart, Warwick, A., Miller, Stuart, D., Smart, Roger, et al. (2006). Advances in Acid Rock Drainage (ARD) Characterisation of Mine Wastes. *Journal of the American Society of Mining & Reclamation (JASMR)*, 2098–2119. doi:10.21000/JASMR06022098.

Stojanova, M., Arbelet, P., Baudin, F., Bouton, N., Caria, G., Pacini, L., Proix, N., Quibel, E., Thin, A., & Barré, P. (2024). Technical note : A validated correction method to quantify organic and inorganic carbon in soils using Rock-Eval ® thermal analysis. *Biogeosciences, 21*, 4229–4237. https://doi.org/10.5194/bg-21-4229-2024.

Svensmark, B. o (2021). Extensions to the Theory of Sampling 1. The extended Gy's formula, the segregation paradox and the fundamental sampling uncertainty (FSU). *Analytica Chimica Acta, 1187*, 339127. doi:10.1016/j.aca.2021.339127.

Tabelin, C., Yoo, K., & Li, J. (2022). *Novel and emerging strategies for sustainable mine tailings and acid mine drainage management* C. Tabelin, Y. Kyoungkeun, & J., Li (Eds.). MDPI. https://doi.org/10.3390/books978-3-0365-2746-8

USEPA. (1992). *Method 1311:* Toxicity characteristic leaching procedure. EPA SW-846: test methods for evaluating solid waste, physical/chemical methods. https://www.epa.gov/hw-sw846/sw-846-test-method-1311-toxicity-characteristic-leaching-procedure

USEPA. (1994). *Method 1312:* Synthetic precipitation leaching procedure. Test methods for evaluating solid waste, physical/chemical methods

USEPA. (2011). *Method 1627: Kinetic test method for the prediction of mine drainage quality*. https://nepis.epa.gov/Exe/ZyPURL.cgi?Dockey=P100IKD9.txt

Valcárcel, M., Cárdenas, S., & Gallego, M. (1999). Sample screening systems in analytical chemistry. *TrAC Trends in Analytical Chemistry, 18*(11), 685–694. doi:10.1016/s0165-9936(99)00167-3.

Vriens, B., Plante, B. ît, Seigneur, N., & Jamieson, H. (2020). Mine waste rock: Insights for sustainable hydrogeochemical management. *Minerals, 10*(9), 728. doi:10.3390/min10090728.

Wangersky, P. J. (1993). Dissolved organic carbon methods: A critical review. *Marine Chemistry, 41*(1-3), 61–74. doi:10.1016/0304-4203(93)90106-X.

Watanabe, K., Badr, E. S., Pan, X., & Achterberg, E. P. (2007). Conversion efficiency of the high-temperature combustion technique for dissolved organic carbon and total dissolved nitrogen analysis. *International Journal of Environmental Analytical Chemistry, 87*(6), 387–399. doi:10.1080/03067310701237023.

Weber, Paul, A., Hughes, Joseph, B., Conner, Liam, B., Lindsay, Phil, Smart, Roger St., C, et al. (2006). *Short-term acid rock drainage characteristics determined by paste pH and kinetic NAG testing: Cypress prospect, New Zealand*. American Society of Mining and Reclamation doi:10.21000/JASMR06022289.

White, W. W., & Jeffers, T. H. (1993). Chemical Predictive Modeling of Acid Mine Drainage from Metallic Sulfide-Bearing Waste Rock. *Environmental Geochemistry of Sulfide Oxidation* (pp. 608–630). Wiley. https://doi.org/10.1021/bk-1994-0550.ch037.

Wilkin, R. T., & Barnes, H. L. (1996). Pyrite formation by reactions of iron monosulfides with dissolved inorganic and organic sulfur species. *Geochimica et Cosmochimica Acta, 60*(21), 4167–4179. https://doi.org/10.1016/S0016-7037(97)81466-4.

Yang, B., Carotta, M. C., Faglia, G., Ferroni, M., Guidi, V., Martinelli, G., & Sberveglieri, G. (1997). Quantification of H2S and NO2 using gas sensor arrays and an artificial neural network. *Sensors and Actuators B: Chemical, 43*(1-3), 235–238. doi:10.1016/s0925-4005(97)00132-9.

Zárate-Gutiérrez, R., Lapidus, G. T., & Morales, R. D. (2012). Aqueous oxidation of galena and pyrite with nitric acid at moderate temperatures. *Hydrometallurgy, 115–116*, 57–63. https://doi.org/10.1016/j.hydromet.2011.12.010.

Zhang, L., Yang, M., Wang, D., Yu, K., Chen, Z., Bing, X., Zhao, P., Bao, Y., & Yin, X. (2024). Accurate determination of high sulfur content in sulfide samples: an optimized ICP-OES method. *Journal of Analytical Science and Technology, 15*(1). doi:10.1186/s40543-024-00463-z.

Zhang, X., Wu, Q., Zhao, D., Du, Z., Zhang, W., Zhu, Q., & Zhang, F. (2026). Geochemical Genesis and Acid Production Potential Assessment of Acid Mine Drainage in Abandoned Mine Sites: An Integrated Study Based on Geochemical Static Tests and Mineralogical Analysis. *Applied Sciences (Switzerland), 16*(1), 1–28. https://doi.org/10.3390/app16010240.

Zhao, W. Z., Lu, B., Yu, J. B., Zhang, B. B., & Zhang, Y. (2020). Determination of sulfur in soils and stream sediments by wavelength dispersive X-ray fluorescence spectrometry. *Microchemical Journal, 156*. Available from www.elsevier.com/inca/publications/store/6/2/0/3/9/1. doi:10.1016/j.microc.2020.104840.

CHAPTER THREE

Conventional treatment methods

Conventional treatment methods for acid mine drainage (AMD) comprise a broad range of physicochemical and biological technologies developed to mitigate the adverse environmental impacts associated with acidic and metal-rich mine effluents. These approaches are primarily aimed at neutralizing acidity, reducing sulfate concentrations, and removing dissolved metals and metalloids prior to discharge into the environment or reuse in mining operations. Conventional AMD remediation technologies are generally classified into active and passive treatment systems, depending on the level of operational control, energy input, and chemical requirements involved (Johnson & Hallberg, 2005; Kefeni et al., 2017). In this Chapter, both active and passive treatment systems have been presented. Additionally, life cycle assessment and economic feasibility has been presented at the end.

3.1 Active treatment systems

Active treatment systems for AMD comprise engineered, continuously operated processes in which chemical reagents—and, in some configurations, biologically mediated reactions—are applied to neutralize acidity and remove dissolved metals. These systems are widely deployed at active mine sites and are specifically designed to ensure high reliability under conditions of elevated flow and extreme acidity. Treatment typically involves controlled dosing with alkaline reagents such as lime ($CaO/Ca(OH)_2$), caustic soda ($NaOH$), or soda ash (Na_2CO_3) to increase pH and promote the precipitation of dissolved metals as hydroxides and related phases. The geochemical basis of this process is well established, with iron and aluminium precipitating over defined pH ranges during neutralization, thereby buffering solution chemistry and enhancing contaminant removal (Nordstrom, 2020). High-density sludge (HDS) systems, a widely adopted modification of conventional active treatment processes, further improve treatment efficiency through the recycling of previously generated sludge into the treatment reactor. The recycled sludge provides nucleation surfaces that enhance precipitation and settling characteristics, resulting in denser

Acid Mine Drainage: Sustainable Mining Practices for Mitigating Environmental Impact
DOI: https://doi.org/10.1016/B978-0-443-45107-2.00002-2

sludge formation, reduced sludge volumes, improved reagent utilization, and lower waste management requirements (e.g. Sukati et al., 2018). In some cases, biological processes are integrated to augment metal removal or provide secondary polishing, although the primary mechanism remains chemically driven precipitation and neutralization (Fuchida et al., 2020). The operation of active treatment systems requires continuous conveyance of AMD to treatment facilities, precise reagent dosing, and the management and disposal of generated sludge, which often contains concentrated metal hydroxides and secondary minerals. For severely contaminated effluents, integrated treatment trains combining chemical neutralization, oxidation, and metal precipitation—sometimes coupled with advanced processes such as membrane filtration—are implemented to meet stringent discharge or reuse standards (Masindi et al., 2017). Numerous physical and chemical methods have been developed for active treatment of AMD. This section focuses on four widely applied approaches: (1) lime dosing and chemical neutralization, (2) aeration and oxidation, (3) ion exchange and adsorption, and (4) reverse osmosis and membrane filtration.

3.1.1 Lime dosing and chemical neutralization

Lime dosing and chemical neutralization are processes of adding controlled amount of lime and other alkaline materials into AMD for dissolved metal precipitation and pH neutralization. Lime dosing and chemical neutralization involve adding alkaline materials into acidic water to raise the pH around 6–9 and precipitate dissolved metals as hydroxides that are subsequently removed as precipitates by gravity settling or by dissolved air flotation (e.g., Aube, 2004; Feng et al., 2000; Kaksonen & Puhakka, 2007). The precipitated metals (in the form of insoluble metal hydroxides) settle out of the acidic contaminated water and get removed in the form of soft, gelatinous sludge, which settles in ponds or tanks. Most of metals precipitate at the pH range of 6–9; thus enough lime is added to raise the pH of water (e.g., Othman et al., 2023). It should be noted however that although all other metals precipitate at that pH range, ferric iron (Fe^{3+}) precipitates around pH 3.5 (Jacobs et al., 2014). While the composition of AMD is mostly influenced by the geology, climate, and mining method(s) practiced within the area, studies have shown that pH, flow rate, water oxidation status, and concentrations of metals in the AMD define the type and size of chemical neutralization systems (Othman et al., 2023).

The choice of chemicals to be applied for AMD treatment depends on both technical and economic factors (Pereira et al., 2020). Technical factors considered here include flow rate, types and concentrations of metals, acidic levels, and receiving stream characteristics. Economic factors include chemical prices, equipment and machinery, labor, financial institutions' interest rates, and treatment duration (e.g., Othman et al., 2023). Popular chemicals used in AMD treatment with their neutralization efficiency (in brackets) include limestone (30%), hydrated lime/calcium hydroxide (90%), soda ash/sodium carbonate (60%), caustic soda/sodium hydroxide (100%), pebble quicklime/calcium oxide (90%), hydrated lime/calcium hydroxide (90%), and ammonia/anhydrous ammonia (100%) (e.g., Kaksonen & Puhakka, 2007; Othman et al., 2023).

Treatment of AMD through lime dosing and chemical neutralization principally relies on two fundamental processes: acid neutralization and metal precipitation. Acid neutralization occurs through reactions between acidic mine water and lime or other alkaline reagents, resulting in the consumption of excess H^+ and a subsequent increase in pH (Feng et al., 2000). As pH rises, the solubility of dissolved metals decreases, promoting their precipitation predominantly as metal hydroxides and oxyhydroxides, although carbonate or sulfate phases may also form under specific geochemical conditions (Feng, 2000; Zhao et al., 2024). Manganese (Mn^{2+}), for example, reacts with OH^- and precipitates as $Mn(OH)_2$ at pH of 8.6–9.9, whereas Fe^{3+} precipitates as $Fe(OH)_3$ at much lower pH ranges typically 1.5–2.9 (Zhao et al., 2024). Other reactions and corresponding precipitates and pHs, as presented by Zhao et al. (2024), are as follows:

$$\text{(i) } Ca^{2+} + 2OH^- \leftrightarrow Ca(OH)_2\downarrow \text{ at } >13.6 \text{ pH} \tag{3.1}$$

$$\text{(ii) } Fe^{2+} + 2OH^- \leftrightarrow Fe(OH)_2\downarrow \text{at } 6.4\text{–}{\sim}9.5 \text{ pH} \tag{3.2}$$

$$\text{(iii) } Mg^{2+} + 2OH^- \leftrightarrow Mg(OH)_2\downarrow \text{ at } >10 \text{ pH} \tag{3.3}$$

$$\text{(iv) } Cu^{2+} + 2OH^- \leftrightarrow Cu(OH)_2\downarrow \text{ at } 6.2\text{–}6.7 \tag{3.4}$$

Regarding acid neutralization, Zhao et al. (2024) performed an investigation on the performance of caustic soda (NaOH) as an alkaline chemical for neutralization of AMD with reference to the Chinese wastewater discharge standard. The acidity of the raw AMD was at pH = 2.16, whereas the standard (GB28661-1012) pH varied between 6 and 7. In the raw AMD, Fe and Mn had relatively higher values, that is, 77.54 and 621.29 mg/L against standard concentration of 5 and 2 mg/L, respectively. Cu, Ca, and Pb concentrations in the raw AMD were in relatively small quantities (6.54, 12.36, and

0.06 mg/L against 0.54, 0, and 1 mg/L, respectively). Zinc, Cd, Cr, Ni, and Ag were not detected in the raw AMD.

As per Zhao et al. (2024), varying dosage quantities of NaOH result in a decrease in metals concentration and pH over time. Zhao et al. (2024) show that optimal results are obtained when 10 mL of AMD is allowed to react with NaOH for 2 hours. Results also indicate that as pH increases following the addition of NaOH in the AMD, removal rate of dissolved metals also increases accordingly. They also show that when NaOH dosage is increased, pH increases from 2.16 to 13.77, and so do removal rate of metals (except for Cu whose removal rate does not change significantly from 1.0 to 2.5 g NaOH dosage). The final metal concentrations as well as the final pH of the treated AMD all show that while at the beginning of the NaOH dosage (1.5 g NaOH dosage), metal concentrations in the AMD were still exceeding the Chinese standard limits, with increase in dosage (at 2.0 g NaOH dosage), the metal concentrations go much lower, meeting the Chinese standard limits. The pH values at final dosage are however too high, highlighting an excess amount of NaOH. Researches show that that the use of lime dosing and chemical neutralization method for AMD treatment is effective (e.g., Othman et al., 2023). However, the use of these chemicals require careful handling as their excess or exposure to environment may be dangerous to human and aquatic life (Hou et al., 2021). Furthermore, chemical treatment is costly, requiring a long time of treatment—some treatment yield high amount of sludge and costly equipment and facilities (Skousen et al., 2019).

3.1.2 Aeration and oxidation

While chemical neutralization provides rapid pH adjustment, aeration enhances treatment efficiency by oxidizing dissolved iron and other metals. Aeration increases the dissolved oxygen through the introduction of air into AMD to facilitate oxidation of dissolved metals, mostly ferrous iron (i.e., Fe^{2+})—thus oxygen in the air combines with dissolved metals in AMD transforming soluble dissolved metals into insoluble forms (i.e., precipitates) (Saha & Sinha, 2016). The importance of aeration lies in the fact that it leads into formation of a more stable form of compound which is ferric hydroxide—a compound that is not stable when exposed to natural precipitation or acidic water (Ackman & Kleinmann, 1984). Oxidation of iron is advantageous because it can ensure treatment efficiency and sludge stability (Aubé, 2004).

$$Fe(OH)_2 + \tfrac{1}{2} H_2O + \tfrac{1}{4} O_2 \leftrightarrow Fe(OH)_3 \text{ (stable ferric hydroxide)} \quad (3.5)$$

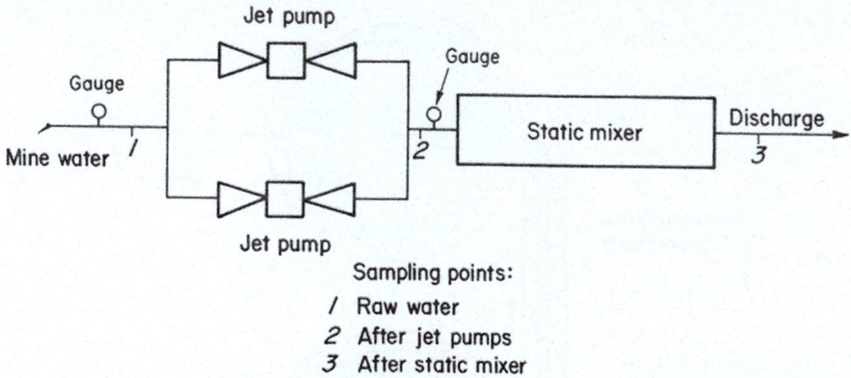

Figure 3.1 In-line aeration system. Sampling points within the in-line system (after Ackman & Kleinmann, 1984). *(From Ackman and Kleinmann (1984)).*

While aeration introduces oxygen into AMD to enhance oxidation, sometimes oxygen rich compounds known as oxidants, such as calcium hypochlorite and sodium hypochlorite (e.g., Hou et al., 2021), may be used instead of undertaking this aeration process. Further, potassium permanganate and hydrogen peroxide are the most effective oxidants, while calcium peroxide seems to be effective in both oxygenation of AMD and neutralization of acidity. In the oxidation process, oxidants are sometimes used to promote metal hydroxide precipitation. Calcium hypochlorite $Ca(OCl)_2$, hydrogen peroxide (H_2O_2), and sodium hypochlorite (NaClO) are strong oxidants, whereas calcium peroxide (CaO_2) is both an oxidant and an acid neutralizer (Jacobs et al., 2014). According to Jacobs et al. (2014), potassium permanganate ($KMnO_4$) is another oxidant that is very effective and commonly used. Researches show that aeration and oxidation methods particularly when combined with chemical neutralization are preferred in treating leachate characterized by high content of dissolved iron, combined methods that result into high efficiency treatment of AMD (Jacobs et al., 2014).

The effect of combining simultaneously aeration and neutralization methods is demonstrated by Ackman and Kleinmann (1984) at a mine site in Greene County, PA, where high concentrations of ferrous iron in mine water had discharge problems. Neutralization by injecting NaOH and aeration were combined into a single step within a built system known as the *in-line aeration* system (ILS) (Fig. 3.1)—the system comprises a jet pump and static mixer both described as aeration and mixing devices, respectively. Furthermore, they designed ponds for collecting treated mine water—the

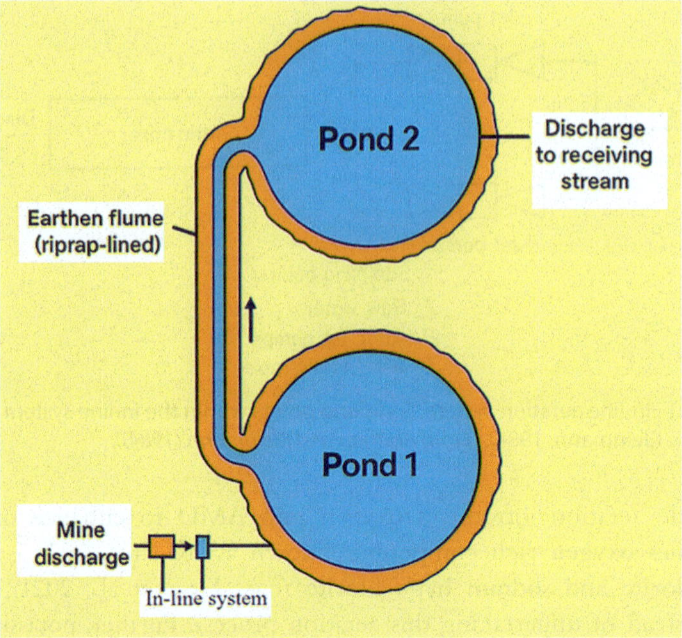

Figure 3.2 Aeration and mixing device. Schematic of ILS and treatment system (after Ackman & Kleinmann, 1984). *ILS*, In-line aeration system. *(From Ackman and Kleinmann (1984)).*

operation that involved collecting three types of water samples (Fig. 3.2). In their design, the water samples were collected as raw water before entering the ILS. The samples were then passed at the jet pump before entering the static mixer, which mixed the samples thoroughly before discharging them (Fig. 3.2).

The *ILS*, as highlighter earlier, is effective in treating AMD, and it has relatively low operation, maintenance, and capital costs. Injection of NaOH at the jet pump of the ILS with simultaneous aeration provides a favorable pH environments for the rapid oxidation of ferrous iron at a relatively short period of time allowing $Fe(OH)_3$ to settle (Ackman & Kleinmann, 1984). According to Ackman and Kleinmann (1984), the iron levels in the discharge are depleted to acceptable levels. For instance, in their experiment, at pH of 7.2, the concentration of Fe^{3+} detected in the discharge water was 1.0 mg/L, while the raw mine water contained 236 mg/L of Fe^{3+} before treatment. Similarly, at a pH of 10.3, the concentration of Fe^{3+} detected in the discharge water was below the detection limit, while the raw mine water contained 137 mg/L Fe^{3+}.

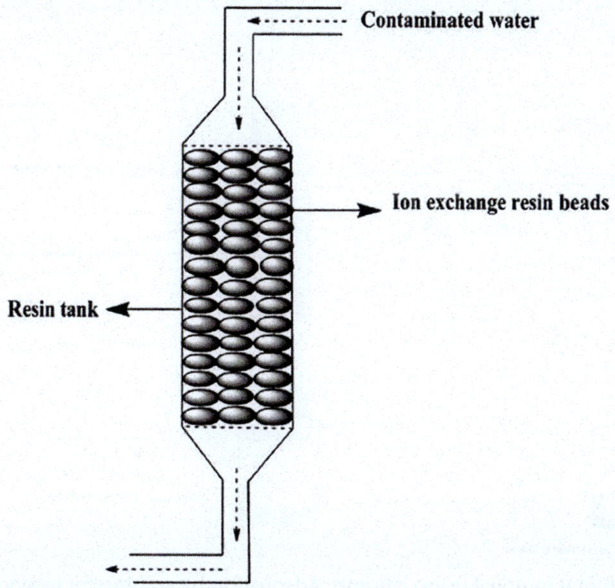

Figure 3.3 Ion exchange for removal of heavy metals (Chakraborty et al., 2020). *(From Chakraborty et al. (2020)).*

3.1.3 Ion exchange and adsorption

Ion exchange is an interchange of ions between a solid resin and an electrolyte solution that involves the exchange of cations with metal ions in AMD (Fig. 3.3). Ion exchange resins are either cation exchangers or anion exchangers (Mulopo, 2022). In AMD treatment, cation exchange resins are commonly used to remove heavy metal cations such as Fe^{2+}, Cu^{2+}, and Zn^{2+} (Feng et al., 2000). These resins contain negatively charged functional groups that attract and bind positively charged metal ions. Ion exchange resins are made up of polymeric structures fitted for various metals, with selectivity for various acidic or chelating functional groups (Chakraborty et al., 2020).

Metals in AMD can be removed by adsorption as well. Adsorption is the accumulation of ions to a surface of a solid material, creating a layer of adsorbate (metal ions) on the surface of the adsorbent (e.g., Maneechakr & Mongkollertlop, 2020). Adsorbent may be derived from different sources such as nonliving biomass including lignin skins and crab, microbial biomass such as bacteria and fungi. Algal bio-mass and other adsorbents altogether may be classified into natural and modified adsorbents (Aprianti et al., 2017; Chakraborty et al., 2020). Batch and column processes are mostly practiced for industrial scale in both adsorption and ion exchange technique

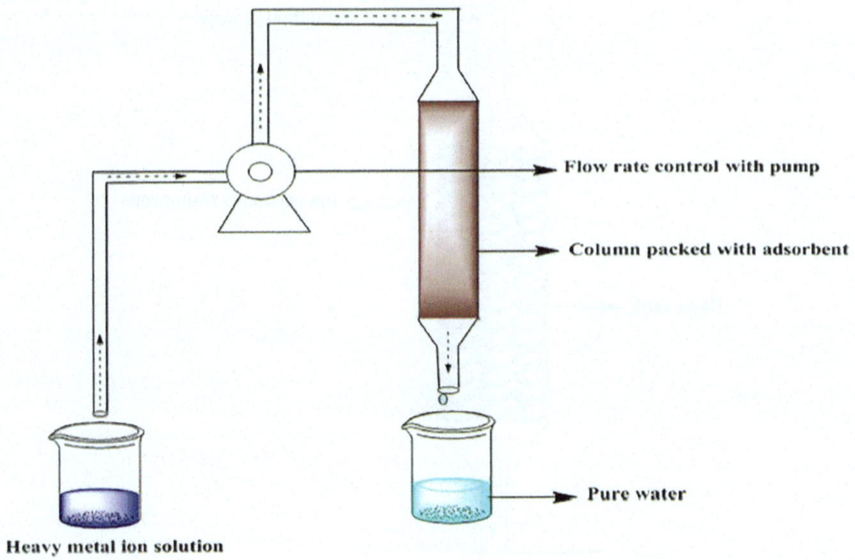

Figure 3.4 Metal removal using column: Adsorption by column for heavy metals removal (Chakraborty et al., 2020). *(From Chakraborty et al. (2020)).*

(Figs. 3.3 and 3.4). During the batch process, the resin and electrolyte solutions are mixed in a tank for a particular period of time so that they can react, and after reaction, the resins are separated from the solution by filtration (Chakraborty et al., 2020). Column is normally used in industrial scale and can be operated through up flow, down flow, or counter flow modes (Sahu & Singh, 2019).

Aprianti et al. (2017) performed an investigation of heavy metal ions on adsorption capacity contained in AMD using activated carbon and clay modified into ceramic adsorbent. The mixture was stirred with powdered carbon and clay until an even mix was obtained and thereafter molded into small balls (5 mm diameter). The evenly mixed and molded balls were then furnaced at 1000°C for 48 hours to create carbon ceramic balls. Forty eight percent NaOH solution was used for the activation mechanism to enhance the adsorption, where carbon ceramic balls were soaked with this 48% of NaOH for 1 hour to ensure unwanted materials deposited inside the pores were removed. Adsorbent samples were created using various activated carbon:clay ratios (Table 3.1).

The adsorption capacity and percentage of adsorption can be determined using Eq. (3.6):

$$qe = \frac{[Co] - [Ct]V}{W} \text{ and } \%qe = \frac{[Co] - [Ct]}{[Co]} \times 100\% \tag{3.6}$$

Table 3.1 Adsorption capacity of adsorbate versus contact time and carbon:clay ratio.

Contact time (minutes)	Adsorption capacity (mg/g)		Carbon: Clay ratio	Adsorption capacity (mg/g)	
	Mn	Al		Mn	Al
0	0.03	0.80	0	0.0	0.0
30	0.23	1.58	2:38	0.02	0.94
60	0.80	2.94	3:37	0.54	2.86
90	2.21	3.59	4:36	1.32	2.94
120	2.39	4.94	5:35	1.75	3.59
150	2.86	5.80	6:34	2.39	4.94
180	3.00	5.93	7:33	2.67	6.39
210	3.07	6.42	8:32	2.71	6.44
240	3.79	6.43	9:31	3.39	6.73
270	4.01	6.57	10:30	3.68	6.87
300	0.03	0.80	11:29	4.11	7.14

Based on Aprianti et al. (2017).

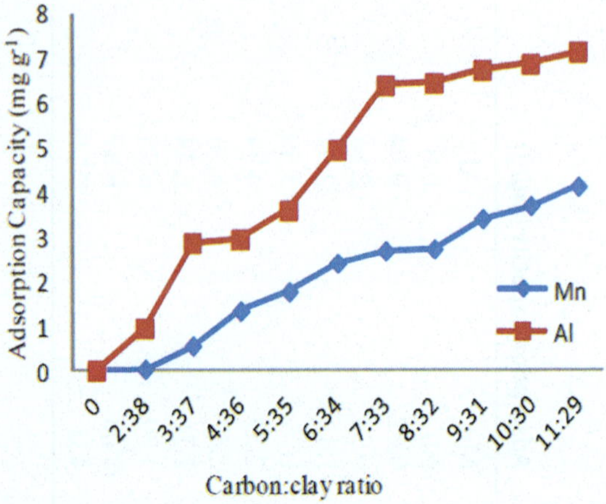

Figure 3.5 Adsorption versus carbon: clay ratio: Metallic ions adsorption capacity with carbon: clay ratio variation on contact time 300 minutes. *(From Aprianti et al. (2017)).*

where *qe* is adsorption capacity (mg/g), *Co* is the initial concentration of the metal in the solution and C*t* is the equilibrium concentration of metal ions in the sample solution (mg/L), *V* is the volume of the testing solution (in liters), and *W* is the weight of the adsorbent in grams (e.g., Chibuzo et al., 2016).

Adsorptive removal of heavy metals for Al and Mn versus pHs of AMD after adsorption show that higher removal of heavy metal ions is achieved at longer contact time between the adsorbent and sample solution (Fig. 3.5; Table 3.1). It may be noted at this stage that contact time varied from 30 to 300 minutes. The adsorption capacity increases from 0.03 to 4.01 mg/g for Mn and 0.8 to 6.57 mg/g for Al (Table 3.1). Additionally, the adsorption capacity was increased along with the carbon:clay ratio and the best optimal contact time and adsorbent ratio was obtained for the treatment of the AMD at 300 minutes and carbon:clay ratio of 11:29 (Table 3.1; Fig. 3.6).

Studies show that the longer the contact time, the higher the pH. This is due to the presence of basic surface functional groups on the activated carbon that reacted with excess hydrogen ions and dissolved metals through ion exchange during adsorption in aqueous solution. As the contact time and carbon: clay ratios are increased, optimal results are obtained for both adsorption capacity of Al and Mn metals and the pH of AMD. The increase in pH can be achieved through three processes: firstly, hydrolysis and the use of activated carbon and NaOH (e.g., Chakraborty et al., 2020). The basic surface functional groups of the activated carbon releases OH ions due

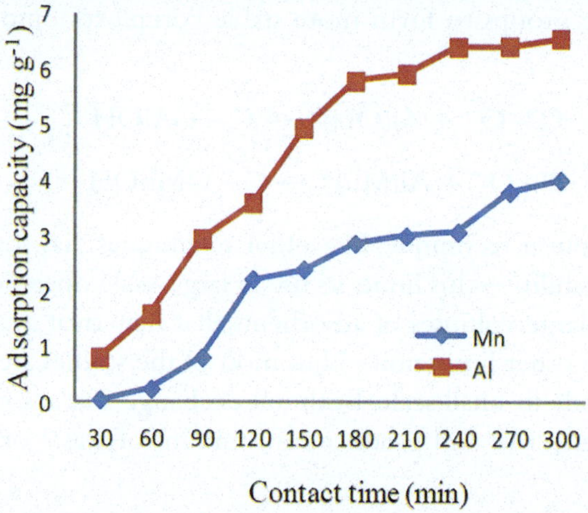

Figure 3.6 Heavy metals adsorption capacity versus contact time variation. The effect of contact time between sample solution and adsorbent on the adsorption capacity. *(From Aprianti et al. (2017)).*

to hydrolysis raising pH, while free H^+ are bound together with $C-O^-$ reducing acidity.

Release of hydroxide ions increasing pH:

$$C-O^- + H_2O \leftrightarrow C-OH + OH^- \tag{3.7}$$

Binding of hydrogen ions reducing acidity:

$$C-O^- + H^+ \leftrightarrow C-OH \tag{3.8}$$

Secondly, ion exchange involves the interaction between dissolved metal cations and reactive surface functional groups present on the exchange material. During this process, exchangeable ions associated with the surface are replaced by dissolved metal ions in solution, resulting in the adsorption and immobilization of heavy metals. Consequently, metal concentrations in the aqueous phase are reduced, while the concurrent release of hydroxyl or other alkaline ions may contribute to a gradual increase in pH. Eqs. (3.9) and (3.10) show how both Al and Mn are captured on the adsorbent surface:

$$2C-O^- + Al^{2+} \leftrightarrow (C-O)_2Al \tag{3.9}$$

$$2C-O^- + Mn^{2+} \leftrightarrow (C-O)_2Mn \tag{3.10}$$

Lastly, metal ions hydrolyze as the pH rises between 4.0 and 7.1 (Aprianti et al., 2017), forming metal hydroxide ions that interact with the activated carbon surface groups to form more stable complexes onto the adsorbent surface(s):

$$C - O^- + Al(OH)^+ \leftrightarrow C - OAlOH \tag{3.11}$$

$$C - O^- + Al(Mn)^+ \leftrightarrow C - OMnOH \tag{3.12}$$

The adsorption technique, like other techniques, has some limitations: (1) unclear feasibility—this limits its use in large-scale operations, (2) limited availability of large volumes of adsorbents that may treat available AMD at large scale is another limitation—thus most of the studies are done in batch or column mode in small scale. Both ion exchange and adsorption method, however, are effective and produce low amount of waste or sludge (Mosai et al., 2024).

3.1.4 Reverse osmosis and membrane filtration

Reverse osmosis is one of the membrane filtration (separation) processes, whereby a selective membrane barrier is used partially or completely to separate a solution (feed) components under certain driving force (pressure). Reverse osmosis involves movement of water under applied driving pressure from high solute concentration to low solute concentration through a semi-permeable membrane (Baker, 2012), typically 0.1–1 nm and operating pressure of 150–1200 PSI (Shannon et al., 2008). Contaminants removed are usually monovalent ions, salts, organics, and metals (Baker, 2012). Further, the membrane that is used in the filtration limits the movement of dissolved ions, organic materials, and other impurities while allowing pure water molecules to pass through (Lee et al., 2011). It is the pressure that creates movement of water across the membrane against the normal concentration gradient to obtain high removal of dissolved ions (Ning, 2002). Reverse osmosis membranes are nonporous with polymer matrix through which water passes via diffusion not through open pores like in microfiltration or ultrafiltration (Baker, 2012; Greenlee et al. 2009). In other words, differences in the solubility and diffusion rate of the solution components in the membrane materials define the reverse osmosis membrane selectivity (Baker, 2012). For effective removal of dissolved solids in AMD treatment, reverse osmosis system requires pretreatment of AMD (Kiswanto & Sundarno, 2020). Pretreatment here refers to initial treatment of the feed water before entering the reverse osmosis plant to reduce the risk of physical damage of the reverse osmosis membrane (Baker, 2012; Kiswanto & Sundarno, 2020). This process involves in-line coagulation, pH adjustment, polymer

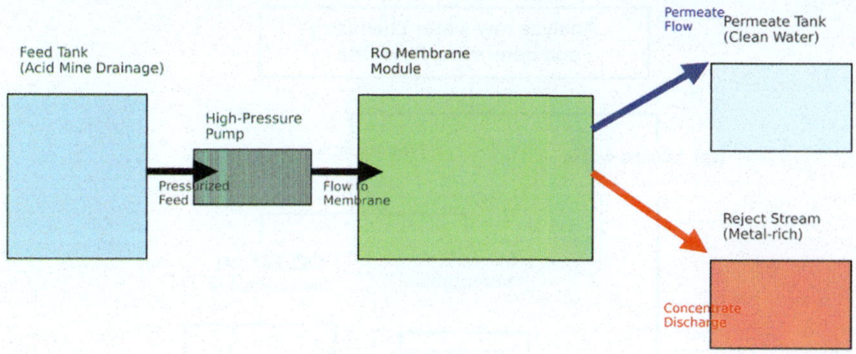

Figure 3.7 RO system. *RO*, Reverse osmosis.

dosing and multimedia filters (Baker, 2012). A fraction of a solution that crosses the membrane is called the permeate (pure/ treated water) and that which remains in the feed (does not cross the membrane) is the reject (concentrate/ contaminated water) (Fig. 3.7). Other membrane filtration processes include ultrafiltration (characterized by 0.01–0.1 μm pore sizes, capable of removing viruses, proteins and colloids at an operating pressure of 15–75 PSI, Greenlee et al., 2009) and microfiltration that is capable of removing bacteria and suspended solids at an operating pressure of 1.5–30 PSI through pore sizes of 0.1–10 μm (Baker, 2012). Nanofiltration occurs in a bit thicker pores (typically 1–10 nm), operating pressure of 60–450 PSI and capable of removing organic and divalent iron materials (Lee et al., 2011).

While membrane filtration have high efficient in contaminant removal and handling toxic materials with steady state operation, it however requires high energy consumption. Additionally, membrane fouling is the biggest challenge in most mining operations when using a reverse osmosis system (Samaei et al., 2020).

3.2 Passive treatment methods

Passive treatment methods refer to those techniques that do not require full time or constant monitoring of chemical reagents as it is the case in active treatment methods discussed in Section 3.1. Thus they may be employed during the mine operation as well as in the post-closure period (Gazea et al., 1996). Passive treatment relies on biological, geochemical, and gravitational processes (e.g., Hedin et al., 1994; Rambabu et al., 2020; Skousen & Ziemkiewicz, 2005; Zipper et al., 2018). The flow chart for

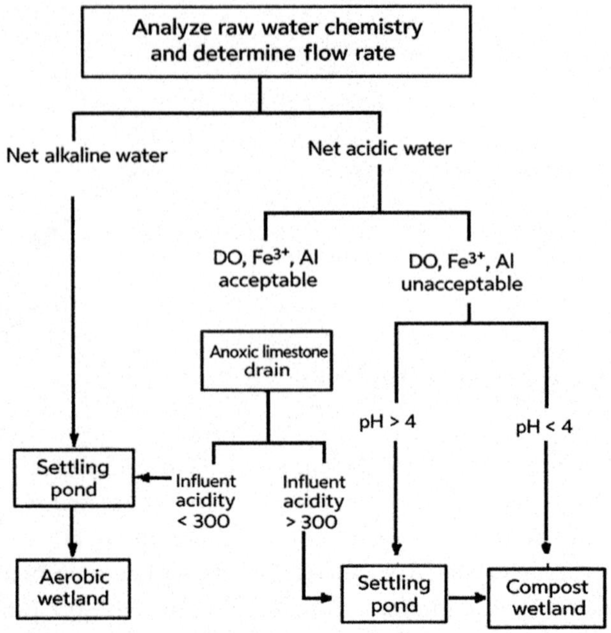

Figure 3.8 Flow chart in passive treatment system: Flow chart showing chemical determinations necessary for the design of passive treatment systems (Gazea et al., 1996). *(From Gazea et al. (1996)).*

passive treatment design is presented in Fig. 3.8. In this section, biological methods are discussed.

3.2.1 Biological approaches

Passive biological approaches for the treatment of AMD, particularly in groundwater remediation applications, are commonly based on the activity of Sulfate-Reducing Bacteria (SRB). These approaches enhance subsurface microbial processes through the injection of concentrated organic substrates or the installation of permeable reactive barriers (PRBs), thereby promoting sulfate reduction, alkalinity generation, and the immobilization of dissolved metals (Rambabu et al., 2020). In this design, infiltration beds, anoxic ponds and modified wetland systems are used to improve surface flow of AMD (Gazea et al., 1996; Kaksonen & Puhakka; 2007; Fig. 3.9). In this section, we shall deal more with constructed wetlands. Constructed wetlands are complex treatment systems that use natural processes to stabilize, sequester, accumulate, degrade, metabolize, and/or mineralize contaminants and have the ability to remove organic and inorganic compounds and suspended solids (ITRC, 2003; Rambabu et al., 2020; Simate & Ndlovu, 2021). Wetlands

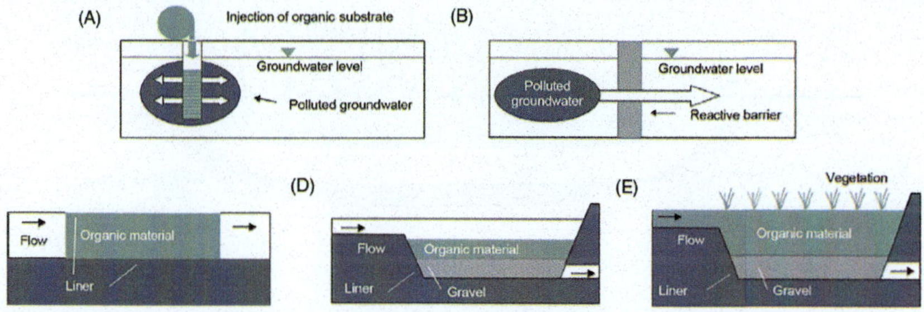

Figure 3.9 Passive biological treatments of AMD: Passive biological treatment applications for AMD-affected ground and surface waters. (A) Substrate injection; (B) reactive permeable barrier; (C) infiltration bed; (D) anoxic pond; (E) anaerobic wetland. *AMD*, Acid mine drainage. *(From Rambabu et al. (2020)).*

are characterized by the existence of water, unique soils that differ from upland soils and the presence of vegetation adapted to saturated conditions (Scholz, 2024). Constructed wetlands treatment usefulness lies in their ability to produce a near-neutral water product which can be readily discharged (Johnson & Hallberg, 2005). According to Simate and Ndlovu (2021), constructed wetlands are characterized by significantly lower total lifetime costs and often lower capital costs than conventional treatment systems—that means, once constructed, they can operate for long periods of time with minimal operations and maintenance. Constructed wetlands are of two types: (1) aerobic constructed wetlands (i.e., wetlands designed and operated in such a way that oxygen is present in the root zone and water column, supporting aerobic microbial processes) and (2) anaerobic constructed wetlands (i.e., wetlands designed with limited or no oxygen in the substrate/water, creating reducing or anaerobic conditions where anaerobic microorganisms dominate (e.g., Jacobs et al., 2024). Other biological approaches include anoxic limestone drains, vertical flow wetlands (also called successive alkalinity-producing systems), limestone and slag leach beds, open limestone channels, and sulfate-reducing bioreactors (Jacobs et al., 2024; Zipper et al., 2010). Temperature, pH, salinity, Eh, and metal concentrations are among the key factors that play a crucial role in determining the effectiveness of passive biological approaches used in the treatment of AMD (Ali et al., 2019; Gazea et al., 1996; Rambabu et al., 2020; Younger., 2000).

3.2.1.1 Principles of aerobic constructed wetlands

Aerobic wetlands are shallow man-made ecosystems (Skousen et al., 1998), typically up to 30 cm (e.g., Simate & Ndlovu, 2021) but operate at depth range of 10–30 cm (e.g., Simate & Ndlovu, 2021). They can also go up to

Figure 3.10 Aerobic wetland: Diagram of a typical aerobic wetland. *(From Gazea et al. (1996)).*

50 cm (Gazea, 1996). They consist of basins or channels, with a relatively impermeable bottom to prevent seepage, soil or another suitable medium to support the vegetation and water at a relatively shallow depth (Gazaea, 1996; Skousen & Ziemkiewicz, 2005) flowing over the soil surface. They are designed to provide an environment for suspended solids and metals to settle and precipitate (e.g., ITCR, 2003). Principally, aerobic wetlands are for net alkaline waters and predominantly for just aeration and precipitation of metals (Simate & Ndlovu, 2021; Skousen & Ziemkiewicz, 2005). Aerobic wetlands are constructed in such a way that they can precipitate metals from water under aerobic conditions, usually in a horizontal flow system (Fig. 3.10). These aerobic wetlands are commonly established in series and contain plants of various species, typically typha and other wetland vegetation species such as *Juncus* and *Scirpus* that encourage more uniform water or leachate flow, help stabilize the substrate, help maintain microbial populations and provide aesthetic qualities to the wetland (Skousen et al., 1998; Skousen & Ziemkiewicz, 2005). In their series, usually lies a small sedimentation basin characterized by no plants in it (Zipper et al., 2018). As per Zipper et al. (2018), waters to be treated flow through the sedimentation basin prior to entering the aerobic wetland, which allows suspended sediments and easily hydrolyzable Fe^{3+} to settle out. Further, the pretreatment basin is designed to extend the aerobic wetland's useful life by accumulating particles that would otherwise settle in the wetland system. As a general rule, aerobic wetlands have a "net alkaline water," they have a typical flow of 10–20 g $Fe/m^2/day$ and 0.5–1 g $Mn/m^2/day$ (Hedin et al., 1994; Skousen & Ziemkiewicz, 2005). Aerobic wetlands provide a substrate and increased water retention times (due to reduced flow rates) for the reaction between influent alkalinity and acidity that is generated from AMD (ITRC, 2015; Simate & Ndlovu, 2021). Metal oxidation, hydrolysis, and oxygen infiltration are typical in these aerobic environment leading to precipitation and physical retention of metals

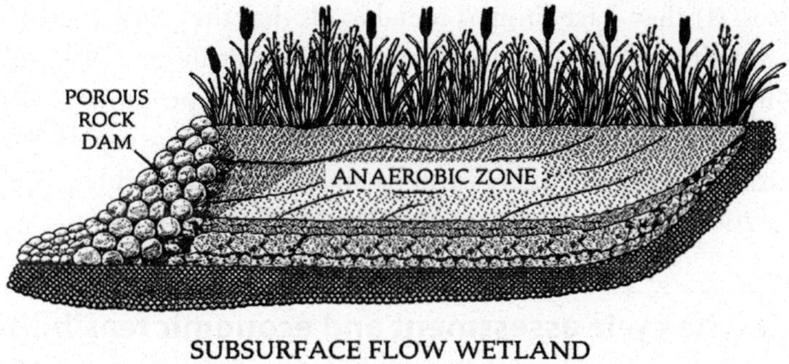

Figure 3.11 Anaerobic wetland: Diagram of a typical anaerobic wetland. *(From Gazea et al. (1996)).*

as oxyhydroxides, hydroxides, and carbonates within the wetland (Simate & Ndlovu, 2021; Taylor et al., 2005). Aerobic wetlands are often used for net alkaline waters and predominantly for just aeration and precipitation of metals (Skousen & Ziemkiewicz, 2005).

3.2.1.2 Principles of anaerobic constructed wetlands

As stated earlier, anaerobic constructed wetlands are designed such that the supply of oxygen in them is limited or completely inhibited (e.g., Doherty et al., 2015; Qin et al., 2026). They are characterized by anaerobic microorganisms that grow and multiply in such oxygen-poor environment (e.g., Jacobs et al., 2024; Fig. 3.11). According to Rambabu et al. (2020), aerobic wetlands target net alkaline waters, whereas the anaerobic wetland involves submerging rich organic substrates, limestone, and sulfate reducing bacteria inoculum for ameliorating acidic metalliferous waters, allowing the reduction of iron and sulfate compounds to be achieved. Chemical and microbial reduction reactions ultimately lead to precipitation of metals and neutralization of acidity (Simate & Ndlovu, 2021). According to Simate and Ndlovu (2021), anaerobic wetlands generate alkalinity through bacterial activity and the use of Fe^{3+} as a terminal electron acceptor. Further, as AMD flows in them (usually in small quantities at a time), the AMD infiltrates through thick, permeable organic material that becomes anaerobic due to high biological oxygen demand. Additionally, since anaerobic wetlands produce alkalinity, their use can be extended to poor quality, net acidic, low pH, high Fe, and high dissolved oxygen AMD.

3.2.1.3 Limitation of wetlands (aerobic and anaerobic)

Although constructed wetlands have a generally low operation costs over long period of time (e.g., Rambabu et al., 2020), they however have their weaknesses: (1) they have limited metal loads that they deal with; hence, adequate pretreatment is required especially when dealing with high volumes and/or highly acidic water; (2) some of the metal precipitates retained in sediments are unstable when exposed to oxygen—thus wetland sediments are supposed to remain largely or permanently submerged (e.g., Johnson & Hallberg, 2005; Simate & Ndlovu, 2021).

3.3 Life cycle assessment and economic feasibility

Life cycle assessment (LCA), the determinant of long-term economic feasibility (e.g., Kwon et al., 2025), is a proven method that can help in the selection of the best available technology in treating AMD by reducing environmental burdens while ensuring efficiency (Masindi et al., 2018; Martínez et al., 2019). The choice and sustainability of remediation systems are the functions of technical and economic factors (Martínez et al., 2019) and sometimes social factors. Integrating economic, technical, as well as social factors into a sustainable LCA provides a complete picture of a project's overall viability and waste material usage strategies, thereby recovering resources from AMD (e.g., Boje et al., 2023). LCA evaluations of environmental impacts have shown that passive AMD treatment systems are often better than active AMD treatment systems, particularly if material transport is minimized (e.g., Moreno-González, et al., 2023). According to Martínez et al. (2019), the construction of the plant for treating AMD initially creates significant environmental impacts, but within 4.5 years, the impacts become negligible. The parameters that are considered in assessing the economic feasibility of a life cycle in treating AMD include climate change (i.e., how the climate has been changed due to construction of a facility for treating AMD), water depletion (i.e., how water has been depleted in the area due to construction of the AMD treatment facility), terrestrial acidification (i.e., how terrestrial acidification has been modified since the construction of such a facility), etc. Martínez et al. (2019) present a detailed account of how different processes (construction, operation, and waste handling) under different timespan scenarios can have impacted the environment due to the construction of a passive AMD treatment facility (Table 3.2) in the Iberian Pyrite Belt—the most significant metallogenic provinces in the world in the SW part of Europe.

In selecting the most appropriate LCA approach, it is essential to align the goal and scope of the study with a thorough inventory analysis and impact assessment (Fig. 3.12). Although the LCA framework applies to

Table 3.2 Impact values of the different processes (construction, C, operation, O, and waste handling, Wh) under different timespan scenarios: 4.5, 20, and 100 years (short, medium, and long term).

Impact category	Unit	4.5 years; short term			20 years; medium term			100 years; medium term		
		C	O	Wh	C	O	Wh	C	O	Wh
Climate change	kg CO_2 eq/m^3	5.95E+00	1.67E+05	3.07E+01	1.34E+00	7.44E+05	1.36E+02	2.68E−01	3.72E+06	6.82E+02
Terrestrial acidification	kg SO_2 eq/m^3	1.24E−02	4.22E+01	6.27E−02	2.80E−03	1.88E+02	2.79E−01	5.60E−04	9.38E+02	1.39E+00
Photochemical oxidant formation	kg NMVOC/m^3	1.88E−02	5.09E+01	1.17E−01	4.23E−03	2.26E+02	5.20E−01	8.46E−04	1.13E+03	2.60E+00
Particulate matter formation	kg PM10 eq/m^3	5.28E−03	1.59E+01	1.95E−02	1.19E−03	7.08E+01	8.68E02	2.38E−04	5.54E+02	4.34E−01
Human toxicity	kg 1,4-DB eq/m^3	2.27E−01	2.34E+01	5.8E−02	5.11E−02	1.04E+02	2.58E−01	1.02E−02	5.20E+02	1.29E+00

(continued on next page)

Terrestrial ecotoxicity	kg 1,4-DB eq/m^3	4.40E−04	1.36E−02	7.70E−07	9.89E−05	6.04E−02	3.42E−06	1.98E−05	3.02E−01	1.71E−05
Freshwater ecotoxicity	kg 1,4-DB eq/m^3	1.24E−04	9.72E−03	8.08E−06	2.78E−05	4.32E−02	3.59E−05	5.57E−06	2.16E−01	1.80E−04
Marine ecotoxicity	kg 1,4-DB eq/m^3	1.17E−03	1.27E−01	5.87E−05	2.63E−04	5.64E−01	2.61E−04	5.25E−05	2.82E+00	1.30E−03
Water depletion	m^3/m^3	5.52E−02	4.48E+02	1.80E−02	1.24E−02	1.99E+03	8.00E−02	2.49E−03	9.96E+03	4.00E−01
Fossil depletion	kg oil eq/m^3	8.87E−01	2.77E+03	9.09E+01	2.00E+01	1.23E+04	4.04E+02	3.99E−02	6.16E+04	2.02E+03
Meta depletion	kg Fe eq/m^3	3.96E−01	0.00E+00	0.00E+00	8.91E−02	0.00E+00	0.00E+00	1.78E−02	0.00E+00	0.00E+00

The short-term period corresponds with the DAS material exhaustion time and replacement needs. The 20-year period is the typical design period used for wastewater treatment plants, while the long-term operation corresponds to the long-lasting effects of AMD. Based on Martínez et al. (2019).

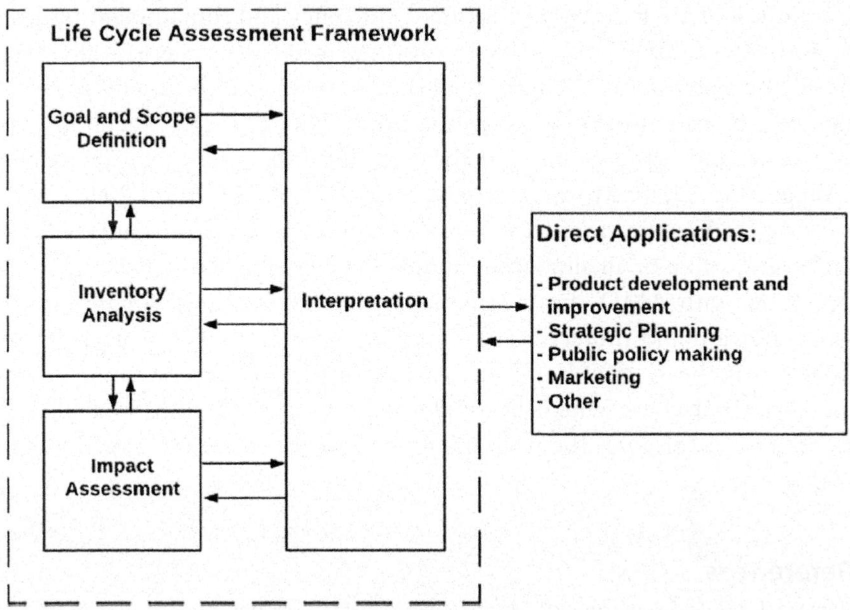

Figure 3.12 Life cycle assessment framework: Stages of an LCA study (Moreau et al., 2020; ISO 14040, 2006). *LCA*, Life cycle assessment. *(From Moreau et al. (2020)).*

high salinity mine water (HSMW) in the same way it does to AMD, its effective use requires specialized datasets and impact categories that reflect the unique environmental characteristics of HSMW. While both AMD and HSMW benefit from LCA-based evaluation, the nature of their environmental burdens, chemical compositions, and treatment pathways differs substantially. For HSMW, the LCA must capture impacts arising from high salt concentrations—including aquatic toxicity, soil salinization, and stress on freshwater resources—which contrast with the heavy-metal loads and low pH conditions typical of AMD (e.g., Moreau et al., 2020).

The HSMW is further characterized by elevated salinity, diverse dissolved inorganic constituents, complex pollutant compositions, and high scaling and corrosive potential, which collectively increase treatment complexity (Feng et al., 2014; Gao et al., 2020). Conventional treatment methods primarily target suspended solids, acidity, and readily precipitable dissolved metals and particulate contaminants. However, the treated effluent frequently retains dissolved salts, persistent contaminants, and other residual constituents that may continue to pose environmental risks and limit opportunities for beneficial water reuse (Pinto et al., 2016; Miller et al., 2021). Although most studies have focused on treatment efficiency and contaminant removal performance, comparatively fewer investigations have systematically evaluated the environmental impacts associated with the treatment processes themselves, including energy consumption, chemical usage, greenhouse gas emissions, and waste generation through life cycle assessment approaches (Wang et al., 2028; Moreno-González et al., 2023; Li et al., 2026). Most existing LCA applications have focused on AMD, with comparatively fewer addressing saline or alkaline mine waters (e.g., Hengen et al., 2014; Li et al., 2026). As with AMD, the treatment of HSMW is critical for preventing salt dispersion and protecting surrounding ecosystems. A major challenge, however, is the scattered and inconsistent nature of HSMW and alkaline mine water datasets, which limits the development of globally applicable LCAs and underscores the need for region-specific databases (e.g., Li et al., 2026).

References

Ackman, T. E. & Kleinmann, R. L. P. (1984). In-line aeration and treatment of acid mine drainage. USBM report of investigations 8868. https://www.researchgate.net/publication/236453827_In-line_aeration_and_treatment_of_acid_mine_drainage.

Ali, H. E. B., Neculita, C. M., Molson, J. W., Maqsoud, A., & Zagury, G. J. (2019). Efficiency of batch biochemical reactors for mine drainage treatment at low temperature and high salinity. *Applied Geochemistry*, (2019). https://doi.org/10.1016/j.apgeochem.2019.01.014.

Aprianti, T., Dawami, B., & Agustina, T. E. (2017). Acid mine drainage treatment using activated carbon ceramic adsorbent in adsorption column. *International Journal on Advanced Science Engineering Information Technology,* 7(4), 1241–1247.

Aubé, B. (2004). *The science of treating acid mine drainage and smelter effluents.* Envirobay Inc From www.envirobay.com.

Baker, R. W. (2012). *Membrane technology and applications* (3rd ed.). John Wiley & Sons, Ltd. https://doi.org/10.1002/9781118359686.

Boje, C., José, V., Menacho, H., Marvuglia, A., Benetto, E., Kubicki, S., Schaubroeck, T., & Gutiérrez, T. N. (2023). A framework using BIM and digital twins in facilitating LCSA for buildings. *Journal of Building Engineering, 76,* 107232. https://doi.org/10.1016/j.jobe.2023.107232.

Chakraborty, R., Asthana, A., Singh, A. K., Jain, B., & Susan, A. H. (2020). Adsorption of heavy metal ions by various low-cost adsorbents: A review. *International Journal of Environmental Analytical Chemistry.* 10.1080/03067319.2020.1722811.

Chizobo, A. A., Asadu, C. O., Ikechukwu, O. M., & Arinze, A. K. (2016). Kinetics and isotherm studies on divalent lead ions adsorption by zeolite solution. *International Journal of Novel Research in Engineering and Science, 3*(1), 49–61. www.noveltyjournals.com.

Doherty, L., Zhao, Y., Zhao, X., Hu, Y., Hao, X., Xu, L., & Liu, R. (2015). A review of a recently emerged technology: Constructed wetland–Microbial fuel cells. *Water Research, 85,* 38–45. https://doi.org/10.1016/j.watres.2015.08.016.

Feng, et al. (2000). Treatment of acid mine water by use of heavy metal precipitation and ion exchange. *Minerals Engineering, 13*(6), 623–642.

Feng, Q., Li, T., Qian, B., Zhou, L., Gao, B., & Yuan, T. (2014). Chemical Characteristics and Utilization of Coal Mine Drainage in China. *Mine Water and the Environment, 33*(3), 276–286. https://doi.org/10.1007/s10230-014-0271-y.

Fuchida, S., Suzuki, K., Kato, T., Kadokura, M., & Tokoro, C. (2020). Understanding the biogeochemical mechanisms of metal removal from acid mine drainage with a subsurface limestone bed at the Motokura Mine, Japan. *Scientific Report, 10*, 20889. https://doi.org/10.1038/s41598-020-78069-92.

Gao, P., Wang, Y., Zhang, Z., Cao, Y., Liu, Z., & Zhang, D. (2020). The general methods of mine water treatment in China. *Desalination and Water Treatment, 202*, 183–205. https://doi.org/10.5004/dwt.2020.26153.

Gazea, B., Adam, K., & Kontopoulos, A. (1996). A review of passive systems for the treatment of acid mine drainage. *Minerals Engineering, 9*(1), 23–42. https://doi.org/10.1016/0892-6875(95)00129-8.

Greenlee, L. F., Lawler, D. F., Freeman, B. D., Marrot, B., & Moulin, P. (2009). Reverse osmosis desalination: water sources, technology, and today's challenges. *Water Research, 43*(9), 2317–2348. doi:10.1016/j.watres.2009.03.010.

Hedin, R. S., Nairn, R. W., & Kleinmann, R. L. P. (1994). *Passive treatment of coal mine drainage.* USDI, Bureau of Mines Information Circular IC 9389.

Hengen, T. J., Squillace, M. K., O'Sullivan, A. D., & Stone, J. J. (2014). Life cycle assessment analysis of active and passive acid mine drainage treatment technologies. Resources, Conservation and Recycling 86, 160–167, ISSN 0921-3449, https://doi.org/10.1016/j.resconrec.2014.01.003.

Hou, Q., Fang, D., Wang, D., Liang, J., & Zhou, L. (2021). Integration of cascaded aeration and neutralization for the treatment of acid mine drainage: Insights into the formation of jarosite. *Hydrometallurgy, 206*. doi:10.1016/j.hydromet.2021.105755.

Interstate Technology and Regulatory Council (ITRC). (2003). Technical and regulatory guidance document for constructed treatment wetlands. Available at http://www.itrcweb.org/GuidanceDocuments/WTLND-1.pdf

Interstate Technology and Regulatory Council (ITRC). (2015). *Mining waste treatment technology selection.* Available at http://www.itrcweb.org/miningwaste-guidance.

ISO 14040. (2026). Environmental management—Life cycle assessment—Principles and framework. International Organization for Standardization.

Jacobs, et al. (2014). *Acid mine drainage, rock drainage and acid sulfate soils: Causes, assessment, prediction, prevention and remediation.* John Wiley & Sons.

Jacobs, J. A., Lehr, J. H., & Testa, S. M. (2024). *Acid mine drainage, rock drainage and acid sulfate soils: Causes, assessment, prediction, prevention and remediation.* John Wiley & Sons.

Johnson, D. B., & Hallberg, K. B. (2005). Acid mine drainage remediation options: A review. *Science of the Total Environment, 338*(1–2), 3–14. doi:10.1016/j.scitotenv.2004.09.002.

Kaksonen, A. H., & Puhakka, J. A. (2007). Sulfate reduction based bioprocesses for the treatment of acid mine drainage and the recovery of metals. *Engineering in Life Sciences.* https://doi.org/10.1002/elsc.200720216.

Kefeni, K. K., Msagati, T. A. M., & Mamba, B. B. (2017). Acid mine drainage: Prevention, treatment options, and resource recovery: A review. *Journal of Cleaner Production, 151*, 475–493. https://doi.org/10.1016/j.jclepro.2017.03.082.

Kiswanto, H., & Sundarno, S. (2020). Treatment of coal mine acid water using NF270 membrane as environmentally friendly technology. *Jurnal Pendidikan IPA Indonesia, 9*(3), 439–450. doi:10.15294/jpii.v9i3.23310.

Kwon, Y., Jang, Y., & An, J. (2025). From short-term profitability to long-term sustainability: exploring life cycle cost in chemical processes. *Cleaner Environmental Systems, 19*, 100358. https://doi.org/10.1016/j.cesys.2025.100358.

Lee, K. P., Arnot, T., & Mattia, D. (2011). A review of reverse osmosis membrane materials for desalination—Development to date and future potential. *Journal of Membrane Science, 370*(1), 1–22. https://doi.org/10.1016/J.MEMSCI.2010.12.036.

Li, M., Tan, X., Fan, J., Chen, H., Zhang, T., Li, C., Wang, S., & Hong, J. (2026). Environmental challenges in advanced treatment of high-salinity mine water: A life cycle assessment perspective. *Environmental Impact Assessment Review, 118*, 108249.

Maneechakr, P., & Mongkollertlop, S. (2020). Investigation on adsorption behaviors of heavy metal ions (Cd^{2+}, Cr^{3+}, Hg^{2+} and Pb^{2+}) through low-cost/active manganese dioxide-modified magnetic biochar derived from palm kernel cake residue. *Journal of Environmental Chemical Engineering, 8*(6), 104467. https://doi.org/10.1016/j.jece.2020.104467.

Martínez, N. M., Basallote, M. D., Meyer, A., Cánovas, C. R., Macías, F., & Schneider, P. (2019). Life cycle assessment of a passive remediation system for acid mine drainage: Towards more sustainable mining activity. *Journal of Cleaner Production*, (211), 1100–1111. https://doi.org/10.1016/j.jclepro.2018.11.224.

Masindi, V., Chatzisymeon, E., Kortidis, I., & Foteinis, S. (2018). Assessing the sustainability of acid mine drainage (AMD) treatment in South Africa. *Journal of the Science of the Total Environment, 635*, 793–802. https://doi.org/10.1016/j.scitotenv.2018.04.108.

Masindi, V., Osman, M. S., & Abu-Mahfouz, A. M. (2017). Integrated treatment of acid mine drainage using BOF slag, lime/soda ash and reverse osmosis (RO): Implication for the production of drinking water. *Desalination, 424*, 45–52. https://doi.org/10.1016/j.desal.2017.10.002.

Miller, K. D., Bentley, M. J., Ryan, J. N., Linden, K. G., Larison, C., Kienzle, B. A., Katz, L. E., Wilson, A. M., Cox, J. T., Kurup, P., et al. (2021). Mine water use, treatment, and reuse in the United States: A look at current industry practices and select case studies. *ACS ES&T Engineering, 1*(11), 1798–1818. https://doi.org/10.1021/acsestengg.1c00244.

Moreau, K., Laamanen, C., Bose, R., Shang, H., & Scott, J. A. (2020). Life cycle assessment to demonstrate how automation improves the sustainability performance of an underground mining operation. *Journal of Sustainable Mining, 19*(3-4). Available at https://doi.org/10.46873/2300-3960.1016.

Moreno-González, R., Macías, F., Meyer, A., Schneider, P., Nieto, J. M., Olías, M., & Cánovas, C. R. (2023). Life cycle assessment of management/valorisation practices for metal-sludge from treatment of acid mine drainage. *Environmental Impact Assessment Review, 99*, 107038. https://doi.org/10.1016/j.eiar.2023.107038.

Mosai, A. K., Ndlovu, G., & Tutu, H. (2024). Improving acid mine drainage treatment by combining treatment technologies: A review. *Science of the Total Environment*, 919. doi:10.1016/j.scitotenv.2024.170806.

Mulopo, J. (2022). Active physical remediation of acid mine drainage: Technologies review and perspectives. *Journal of Ecological Engineering, 23*(6), 148–163. doi:10.12911/22998993/147409.

Ning, R. Y. (2002). Arsenic removal by reverse osmosis. *Desalination, 143*, 237–241. https://doi.org/10.1016/S0011-9164(02)00262-X.

Nordstrom, D. K. (2020). Geochemical modeling of iron and aluminum precipitation during mixing and neutralization of acid mine drainage. *Minerals, 10*(6/547), 12. https://doi.org/10.3390/min10060547.

Othman, A., Mamata, C. R., Ibrahim, I., & Rozib, M. S. M. (2023). The mixture of limes in acid mine drainage. *Malaysian Journal of Fundamental and Applied Sciences, 19*, 668–678. doi:10.11113/mjfas.v19n4.2759.

Pereira, et al. (2020). Acid mine drainage (AMD) treatment by neutralization: Evaluation of physical-chemical performance and ecotoxicological effects on zebrafish (Danio rerio) development. *Chemosphere*, 253. doi:10.1016/j.chemosphere.2020.126665.

Pinto, P. X., Al-Abed, S. R., Balz, D. A., Butler, B. A., Landy, R. B., & Smith, S. J. (2016). Bench-scale and pilot-scale treatment technologies for the removal of total dissolved solids from coal mine water: A review. *Mine Water and the Environment, 35*(1), 94–112. https://doi.org/10.1007/s10230-015-0351-7.

Qin, H., Sun, Y., Liu, T., Liu, X., Nie, W., Qu, H., Li, J., & Chen, Y. (2026). Hematite-facilitated anaerobic oxidation of organics: Novel strategy to alleviate bioclogging in constructed wetlands. *Bioresource Technology, 439*, 133290. https://doi.org/10.1016/j.biortech.2025.133290.

Rambabu, K., Banat, F., Pham, Q. M., Ho, S-H, Ren, N-Q, & Show, P. L. (2020). Biological remediation of acid mine drainage: Review of past trends and current outlook. *Environmental Science and Ecotechnology, 2*, 100024. https://doi.org/10.1016/j.ese.2020.100024.
Saha, S., & Sinha, A. (2016). A review on different active treatment methods for acid mine drainage. *Journal of Basic and Applied Engineering Research, 3*(8), 676–680.
Sahu, O., & Singh, N. (2019). Significance of bio-adsorption process on textile industry wastewater. In Shahid-ul-Islam, & B. S. Butola (Eds.), *The Textile Institute Book Series, The impact and prospects of green chemistry for textile technology* (pp. 367–416). Woodhead Publishing ISBN 9780081024911. https://doi.org/10.1016/B978-0-08-102491-1.00013-7 .
Samaei, S. M., Gato-Trinidad, S., Altaee, A., et al. (2020). Performance evaluation of reverse osmosis process in the post-treatment of mining wastewaters: Case study of Costerfield mining operations, Victoria, Australia. *Journal of Water Process Engineering, 34*. doi:10.1016/j.jwpe.2019.101116.
Scholz, M. (2024). *Constructed wetlands. Wetlands for water pollution control* (3rd ed, pp. 139–158). Elsevier. https://www.sciencedirect.com/science/article/pii/B9780443138386000203. doi:10.1016/B978-0-443-13838-6.00020-3.
Shannon, M., Bohn, P., Elimelech, M., et al. (2008). Science and technology for water purification in the coming decades. *Nature, 452*, 301–310. doi:10.1038/nature06599.
Simate, G. S., & Ndlovu, S. (2021). Acid mine drainage from waste to resources. CRC Press. Taylor & Francis Group, LLC: 332 p.
Skousen, J., & Ziemkiewicz, P. (2005). *Performance of 116 passive treatment systems for Acid Mine Drainage*. ASMR.
Skousen, J., Rose, A., Geidel, G., Foreman, J., Evans, R., Hellier, W., et al. (1998). *Handbook of technologies for avoidance and remediation of Acid Mine Drainage* p. 132. The National Mine Land Reclamation Center.
Skousen, J. G., Ziemkiewicz, P. F., & McDonald, L. M. (2019). Acid mine drainage formation, control and treatment: Approaches and strategies. *The Extractive Industries and Society*, 241–249. doi:10.1016/j.exis.2018.09.008.
Sukati, B. H., De Jager, P. C., Annandale, J. G., & Tanner, P. D. (2018). The hazardous status of high density sludge from acid mine drainage neutralization. *Sustainability, 10*(11), 1–19. art. 4185 http://hdl.handle.net/2263/70949 .
Taylor, J., Pape, S., & Murphy, N. (2005). A summary of passive and active treatment technologies for acid and metalliferous drainage (AMD). Australian Centre for Minerals Extension and Research (ACMER). Fifth Australian Workshop on Acid Drainage, 29-31 August 2005, Fremantle, Australia. Earth Systems PTY Ltd Australian Business No. 29 006 227 532 Environment - Water - Sustainability.
Wang, Y., Tavakkoli, S., Khanna, V., Vidic, R. D., & Gilbertson, L. M. (2018). Life Cycle Impact and Benefit Trade-Offs of a Produced Water and Abandoned Mine Drainage Cotreatment Process. *Environmental Science & Technology, 52*(23), 13995–14005. https://doi.org/10.1021/acs.est.8b03773.
Younger, P. (2000). The adoption and adaptation of passive treatment technologies for mine waters in the United Kingdom. *Mine Water and the Environment, 19*, 84–97.
Zhao, P., Zhang, R., & Hu, M. (2024). Alkaline chemical neutralization to treat acid mine drainage with high concentrations of iron and manganese. *Water, 16*, 821. doi:10.3390/w16060821.
Zipper, C., Skousen, J., & Jage, C. (2010). Passive treatment of acid-mine drainage. Reclamation guidelines for surface mined lands. Virginia Cooperative Extension Publication 460-133. Powell River Project.
Zipper, C., Skousen, J., & Jage, C. (2018). Passive treatment of acid-mine drainage. *Virginia Cooperative Extension, Virginia Tech*. PUBLICATION 460-133. VT/0218/460-133 (CSES-216P). www.ext.vt.edu. https://ext.vt.edu/content/dam/pubs_ext_vt_edu/460/460-133/CSES-216.pdf.

CHAPTER FOUR

Non-conventional treatment methods

4.1 Non-conventional treatment methods

Non-conventional treatment methods for AMD refer to approaches that extend beyond traditional chemical neutralization and settling processes. These methods have been developed to address the limitations of conventional treatments, including high sludge production, long-term operational costs, limited metal recovery, and increasing environmental sustainability requirements. Many nonconventional approaches aim to improve metal removal efficiency, reduce chemical consumption, and enhance treatment reliability under complex site conditions (e.g., Muedi et al., 2025). This chapter reviews nonconventional AMD treatment methods classified into active, integrated, and passive systems. Active methods include biologically mediated processes, electrochemical techniques, nanotechnology-based treatments, membrane filtration, ion exchange, and artificial intelligence–assisted monitoring and control systems, which typically offer rapid treatment and high removal efficiencies but require greater operational input (e.g., Eniola et al., 2024). Integrated treatment systems combine chemical, biological, and physical processes to improve overall performance and flexibility. Passive methods rely on natural processes such as phytoremediation, algal treatment, constructed wetlands, and biochar-based systems, offering lower cost and long-term sustainability, albeit at slower treatment rates. Collectively, these approaches provide a diverse set of tools for AMD management, with their suitability governed by site-specific factors such as drainage chemistry, flow conditions, climate, land availability, and long-term mine closure objectives.

4.1.1 Active methods

Traditionally, active treatment of AMD relies on the addition of alkaline reagents such as NaOH, Na_2CO_3, NH_3, lime, and limestone to neutralize acidity and raise pH (Trumm, 2010). To increase efficiency in treating AMD, addition of coagulants and flocculants to accelerate the rate of precipitation

Acid Mine Drainage: Sustainable Mining Practices for Mitigating Environmental Impact
DOI: https://doi.org/10.1016/B978-0-443-45107-2.00003-4

of heavy metals in settling ponds is commonly employed (Rose, 2010; Roy & Bhattacharya, 2015). One of the principal limitations associated with traditional active treatment systems is the requirement for continuous and often costly operation and maintenance, (2) generation of secondary pollutants, for example, massive sludge production, brines, CO_2, and H_2S (Masindi et al., 2022). Following the advancement of science and technology however, new emerging active methods such as the use of microorganism are being employed to increase efficiency and rate of treatment process (Adetunji & Erasmus, 2025; Du et al., 2022; Rambabu, 2020). Nanotechnology, electrochemical, membrane filtration (Beauclair & Ambushe, 2024; Mulopo, 2022; Roy & Bhattacharya, 2015), and AI-driven systems are gaining greater reputation as active techniques for treatment of AMD (Mardonova et al., 2025).

4.1.1.1 *Microbial and biotechnological approaches*

Microbial and biotechnological AMD treatment uses microorganisms, mainly bacteria but also fungi and algae, to convert hazardous AMD constituents into stable, non-toxic compounds such as CO_2, H_2O, and biomass (Adetunji & Erasmus, 2025; Du, 2022; Rambabu, 2020). These approaches can be categorized into **active** and **passive** treatments, based on the degree of environmental control, energy input, and operational maintenance involved (Trumm, 2010).

In active microbial treatment, microorganisms are introduced into engineered systems where optimal environmental conditions are maintained to enhance their metabolic activity. These systems are tightly managed and require continuous monitoring of key parameters such as pH, temperature, redox potential, salinity, and nutrient availability. Active treatment often involves the cultivation of specific strains of sulfate-reducing bacteria (SRB), acidophilic SRB, and iron-oxidizing bacteria under controlled conditions. Generally, for optimum growth of SRB, the optimum temperature is about 40°C and 5–9 pH (Rambabu, 2020), while for iron-oxidizing bacteria growth, the ideal temperature is 30°C–40°C (Mao et al., 2024). Organic compounds such as ethanol, lactate, glycogen, acetate, or formate are normally supplied as electron donors to stimulate microbial activity and facilitate AMD treatment (Adetunji & Erasmus, 2025; Du, Bogush, Mašek, Purton, & Campos, 2022; Zhao, 2018).

In these systems, SRB reduce SO_4^{2-} to H_2S using organic matter as an electron donor (Eqs. 4.1–4.4). The H_2S subsequently reacts with dissolved metal ions (M^{2+}) to form insoluble metal sulfides $MS_{(s)}$, which precipitate

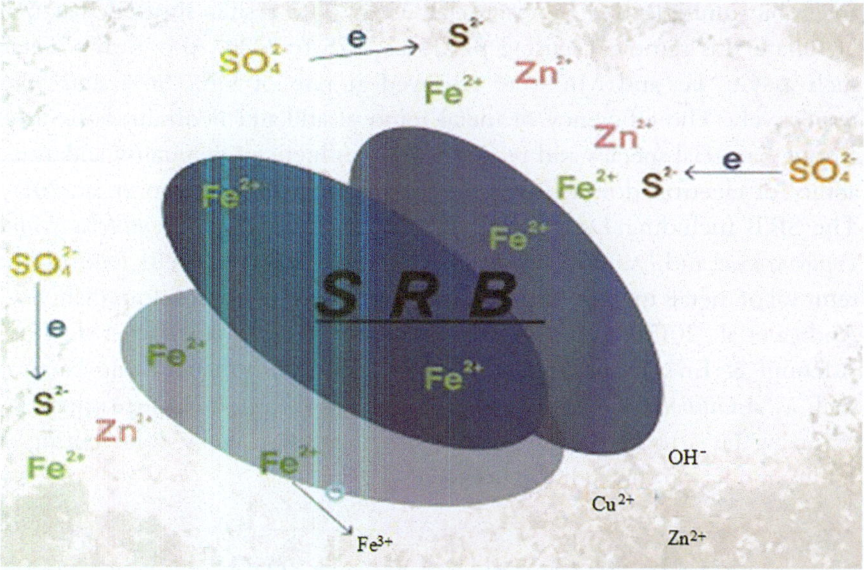

Figure 4.1 Mechanism of SRB. *SRB*, Sulfate-reducing bacteria. *(From Du et al. (2022)).*

out of solution. Simultaneously, the production of HCO_3^- helps to neutralize acidity by reacting with H^+ to form CO_2 and H_2O. These biochemical reactions are represented as follows (Papirio et al., 2013; Panda, 2016):

$$SO_4^{2-} + 4H_2 + H^+ \rightarrow HS^- + 4H_2O \quad (4.1)$$

$$\text{Organic carbon} + SO_4^{2-} \rightarrow HS^- + HCO_3^- \quad (4.2)$$

$$M^{2+} + HS^- \rightarrow MS\downarrow + H^+ \quad (4.3)$$

$$HCO_3^- + H^+ \rightarrow CO_2 + H_2O \quad (4.4)$$

Depending on the factors such as species of SRB, type of metal and pH, some of the SRB remove metals in AMD through bio-precipitation and extracellular chelation (Fig. 4.1). The mechanism of metals removal from AMD significantly depends on the type of metal removed. For instance, Cu is removed as a hydroxide precipitate by a chelating compound while Fe and Zn can be removed by bioprecipitation as metal-sulfide precipitates (Du et al., 2022). Zhao (2018) has shown that at pH above 8 most of the metal hydroxides precipitate. For instance, Fe precipitates at a pH range of 9.2–9.5, Zn precipitates at a pH 8.3–8.5, and Cu precipitates at a pH 7.5–8.0.

Bai et al. (2013) worked on the use of SRB enhanced with Fe° for the treatment of heavy metals and pH neutralization in AMD using upflow anaerobic multiple bed bioreactor at 25°C. The results showed that 61% of sulfate was removed, raising pH from 2.75 to 6.20. Also, heavy metals such as Cu, Fe, and Mn were removed at rate of 99%, 86%, and 53%, respectively. The efficiency of metal removal and pH neutralization varies among bacterial species and is significantly influenced by quality and availability of electron donors, pH, and temperature (Nancucheo et al., 2017). The SRB including *Desulfosporosinus, Thermodesulfobium, Desulfitobacterium, Peptococcaceae* and *Desulfotomaculum* have demonstrated high efficiency in the removal of metals under controlled laboratory and field conditions (Sanchez-Andrea et al., 2015; Phyo et al., 2020; Dev et al., 2021; Nobahar et al., 2024; Adetunji & Erasmus, 2025). In addition to SRB, iron-oxidizing bacteria such as *Acidithiobacillus ferrooxidans* play a vital role in AMD treatment by oxidizing Fe^{2+} to Fe^{3+} while consuming protons, thereby contributing to acidity reduction through the reaction:

$$4Fe^{2+} + O_2 + 4H^+ \rightarrow 4Fe^{3+} + 2H_2O \tag{4.5}$$

This process is effective under optimized temperature, typically between 25°C and 35°C and pH 2.0 and 4.0 (Mao et al., 2024). The high level of control of parameters in active treatment systems leads to rapid removal of metals and high metal efficiency. Typical reported values of Fe, Cu and Cr removal in such systems can reach approximately 93%, 82%, and 87%, respectively (Wang, 2021; Dong et al., 2023). In some systems pH control and treatment stability is achieved by incorporating alkaline agents such as lime, soda ash, limestone, or caustic soda (Kefeni, 2017). This is associated with higher costs, energy demands, and operational complexity.

Many studies have been conducted on the use of SRB for the removal of heavy metals and adjusting pH toward neutrality in AMD (e.g. Wang et al., 2021). In their work, they showed removal efficiency of 86% for Fe and 60% for sulfate, which precipitated out as mineral jarosite. Similar studies were done by Dong et al. (2023) where SRB was used in batch remediation to remove heavy metals from AMD. The removal efficiencies of SRB for different metals were 92.16%–93.17% for Fe at initial concentrations of 0–300 mg/L, 83.51%–87.21% for Cr at an initial concentration of 100–500 mg/L, 79%–86% for Mn at an initial concentration of 0–40 mg/L, and 79.9%–81.80% for Cu at an initial concentration of 0–20 mg/L. These results suggest that metal ion concentration influences the inhibition of SRB activity, with inhibition being particularly low for Cr.

4.1.1.2 *Advanced technologies*

In this subchapter, nanotechnology, electrochemical treatments, membrane filtration, and AI-driven systems are discussed as part of the advanced technologies.

4.1.1.2.1 Nanotechnology

Nanotechnology utilizes engineered nanoparticles to remove heavy metals and, in many cases, to reduce acidity by adjusting pH (Del Prado-Audelo et al., 2021; Mulopo, 2022; Beauclair & Ambushe, 2024; Olawade et al., 2024). Nanoparticles remove metals from AMD through ion exchange, precipitation, and nanofiltration (Boyd, 2013; Del Prado-Audelo et al., 2021; Olawade et al., 2024), in which case a metal is fixed by a functional group of either carboxyl, hydroxyl, or amino group at the surface area of a particular nanoparticle (Dlamini et al., 2021). Nanoparticles such as carbon-, polymer-, and silica-based are commonly used (Del Prado-Audelo et al., 2021). Nanoparticles may be nano-clay, modified silica, Fe–Cu, polyethyleneimine, and Mg-oxide (Olawade et al., 2024). Nanoparticles may be synthesized either chemically or biologically, though biological nanoparticles are preferred due to their friendliness to the environment (Adetunji & Erasmus, 2025). Biologically synthesized nanoparticles are also nontoxic and characterized by large surface areas (Kapoor, 2021) that facilitate increased rate of metal removal from AMD (Kapoor, 2021; Loschinina, 2023). Additionally, the cost to make biologically synthesized nanoparticles is relatively low (Kapoor et al., 2021). Fe nanoparticles, being magnetic, have advantages of binding heavy metals in AMD and the bound metals can later be released by magnetic separation (Simate, 2021). For instance, at optimal pH of 2.5, 6.5, and 8.5, Hu et al. (2006) demonstrate that synthetic Fe-nanoparticle maghemites can be used to remove Cr, Cu, and Ni, respectively. The limitation of using nanoparticles, however, is that while nanoparticles are compatible with other treatment methods and are environmentally friendly, their efficiency is limited to parameters such as temperature and pH. High temperature and pressure beyond optimum levels can cause denaturation of functional groups of nanoparticles (Adetunji & Erasmus, 2025).

Most of studies on nanotechnology focus on the treatment of heavy metals from wastewater (Del Prado-Audelo et al., 2021; Dlamini et al., 2021; Olawde, 2024). Dlamini et al. (2021) studied the use of synthesized bimetallic Fe–Cu nanoparticles for treatment of heavy metals and pollutants from coal mine wastewater. The synthesized Fe–Cu bimetallic nanoparticle was produced by combining two metals of Fe and Cu to form bimetallic nanoparticle which was enhanced with bioflocculant containing functional group of carboxyl, hydroxyl, and amino to accelerate the binding rate of metal pollutant from wastewater. The outcome demonstrated the successful removal of phosphates, sulfates, calcium, and pH neutralization.

4.1.1.2.2 Electrochemical technique

Another advanced method for treating AMD is through electrochemical treatment—it uses electrochemical cells (Bejan & Bunce, 2015; Brewster, 2020).In this approach, metals from AMD are removed without addition of alkaline chemicals for production of sludge and subsequently neutralizing pH (e.g., Brewster, 2020). This is achieved by passingelectric current between anode and cathode in a built-up electrochemical cell (Mulopo, 2022; Simate, 2021; Tran, 2017). The electric current initiates redox reaction through electrocoagulation, electroflotation, and electrodeposition (Bejan & Bunce, 2015; Mulopo, 2022; Simate, 2021). The reaction that occurs at the cathode (i.e., reduction) consumes H^+, thus neutralizing pH (Eq. 4.6).

$$H^+ + e^- \rightarrow 1/2\ H_2 \tag{4.6}$$

Metals, such as Ni and Cu, concentrate at the cathode by electrodeposition (Brewster et al., 2020), whereas at the anode, oxidation occurs (Bejan & Bunce, 2015; Simate, 2021). As the pH rises due to the consumption of H^+, solubility of dissolved metals decreases and precipitates out as hydroxides (Brewster, 2018; Chartrand & Bunce, 2003). Khelifa et al. (2005) used electroflotation to reduce the concentration of Cu and Ni from AMD by floating the metals on the surface of hydrogen and oxygen gas bubbles and managed to achieve 98%–99% removal of Cu and Ni at optimum current density, pH, and initial heavy metal concentration. This process uses sacrificial anode such as Fe and Al to generate coagulants where anode generates metal ions and cathode generates hydrogen gas (H_2g). In other words, electroflotation uses H_2 and O_2 gases generated by cathode and anode reaction to bind heavy metals and float as bubbles on water—thereby removing metals and neutralizing pH (Bejan & Bunce, 2015) such that:

$$\text{Anode: } Fe \rightarrow Fe^{2+} + 2e^- \tag{4.7}$$

$$\text{Cathode: } H^+ + e^- \rightarrow 1/3H_2 \text{ or } H_2O + e^- \rightarrow 1/2H_2 + OH^- \tag{4.8}$$

Studies have shown that, despite the fact that electrochemical treatment is relatively expensive as it consumes more energy, yet, electrochemical cell has proven success for the treatment of metals from AMD and neutralizing pH (Brewster et al., 2020). Electrochemical treatment has greater metal removal ranging between 40% and 90% depending on the metal type. Further, this process is environmentally friendly with less amount of sludge formation (Bejan & Bunce, 2015).

4.1.1.2.3 Membrane filtration

This is a process of separating metals and suspended solids in AMD by using membrane as a filter (Beauclair & Ambushe, 2024). Based on the pressure difference, chemical potential difference, electrical difference, and temperature difference four categories of pressure-driven membrane filtration processes can be made. This includes microfiltration, nanofiltration, reverse osmosis and ultrafiltration (Abdullah et al., 2019; Abhanga, 2013). Microfiltration is characterized by membrane thickness of 1–150 μm, pore size of 0.02–4 μm and operational pressure of 2 bar (Mortazavi, 2008; van der Bruggen, 2003). Nanofiltration membrane with 1–150 μm membrane thickness, pore size of less than 0.002 μm is made from cellulose acetate (e.g., Lau et al., 2019). Having the operating pressure range between 5 and 35 bar is highly efficient in removing monovalent metals in AMD through diffusion or exclusion process (Mortazavi, 2008; van der Bruggen, 2003).

Reverse osmosis, also part of membrane filtration, is characterized by a 1–15 μm membrane thickness made from cellulose acetate with 0.02 μm pore size and an operating pressure of 15–150 bar (Abdullah et al., 2019). It is commonly used for water purification and desalination. Ultrafiltration has 1–250 μm membrane thickness and 0.02–0.2 μm pore size range that can separate heavy metals from AMD through sieving at an operating pressure of 1–10 bar. Ultrafiltration can remove the heavy metal and dissolved solids that are left by microfiltration while nanofiltration removes the heavy metal and dissolved solids left by ultrafiltration (Fig. 4.2; Abdullah et al., 2019). Regarding ultrafiltration, micellar enhanced filtration or polymer enhanced filtration can be introduced in membrane filtration to increase efficiency in removing metals such as Cd, Cu, Ni, Pb, and Zn (Agboola, 2019).

Unlike the efficiency of electrochemical method in treating AMD, membrane filtration has even higher efficiency (Rambabu et al., 2020). For instance, nanofiltration has an efficiency of 93%–98%, and reverse osmosis has 99% efficiency (Rambabu et al., 2020). The other advantage of membrane filtration over other remediation approaches is that it can save 15% energy consumption (Gyu, 2017). It also operates at room temperature with the ability to remove both organic and inorganic substances (Agboola, 2019). The challenge associated with membrane filtration however is keeping the flow rate intact in relation to hydraulic pressure needed to push heavy metals against the membrane (Ambiado et al., 2017).

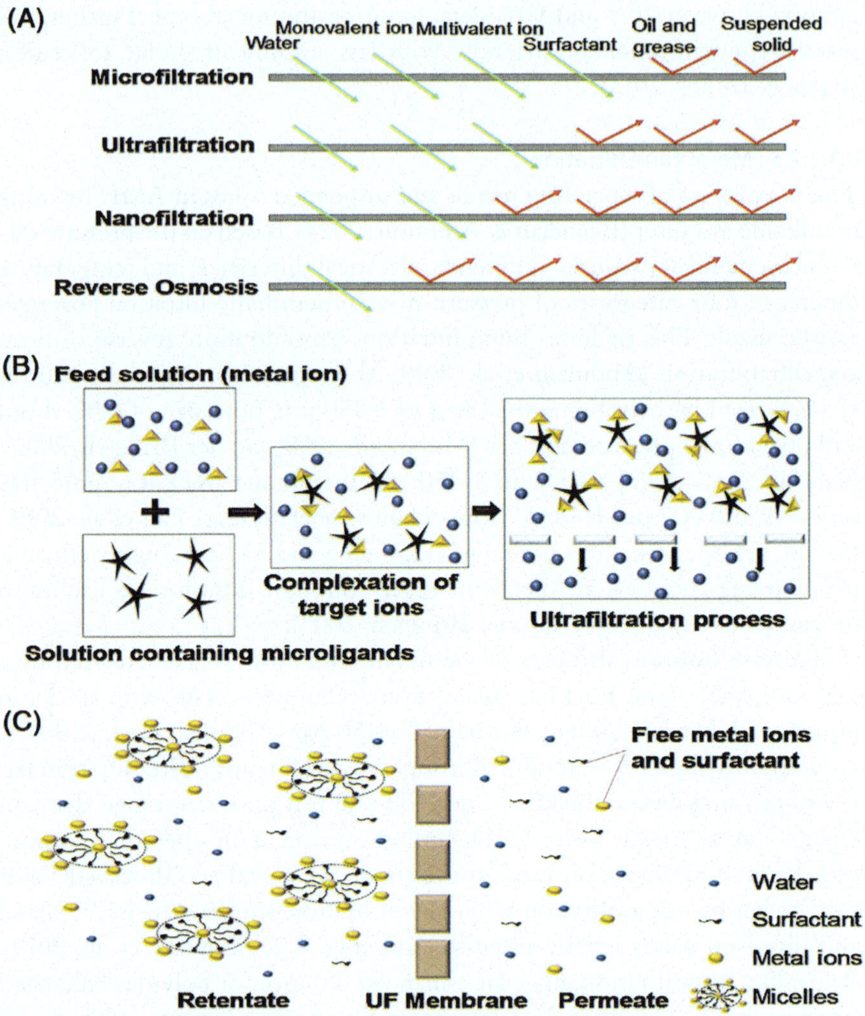

Figure 4.2 Membrane filtration. (A) Rejection capabilities of different membranes against wide range of solutes, (B) separation of heavy metals complexes by Complexation-enhanced ultrafiltration, and (C) separation of heavy metals by Micellar-enhanced ultrafiltration. *(From Abdullah et al. (2019)).*

4.1.1.2.4 Ion exchange

The ion exchange method can be classified into cationic, anionic, and chelating ion exchange (Dinardo et al., 1991). Ion exchanger or ion exchanging resins use anions and cations within a given AMD by allowing them to exchange their positions (Feng, Aldrich, & Tan, 2000; Mulopo, 2022).

The ion exchange may utilize natural or synthetic ion exchanger (resin) to remove metals and neutralize pH (Dlamini et al., 2019; Mulopo, 2022). The common ion exchanger used are zeolite, bentonite, and magnesite (Beauclair et al., 2024). Anionic exchange binds negatively charged and amphoteric metals such as arsenic, whereas cationic exchanger, on the other hand, are made up of carboxylic or sulfonic functional groups—they bind multimetals of polyvalent charges such as Mg, Fe, and Ca (Dinardo et al., 1991). Chelating ion exchanger resin is highly selective for specific metals such as Cu (Dorfner, 1991; Gaikwad, 2009). Ion exchange occurs in two phases, namely, adsorption phase and regeneration phase (e.g., Li & Yu, 2014). In adsorption phase, AMD is channeled across resin exchanger bed, whereby resin materials such as zeolite and bentonite bind the metals onto their surfaces, thereby removing metals and neutralize pH (Beauclair et al., 2024; Mulopo, 2022). During the *loading/adsorption phase*, metal binding occurs on the surface of the exchange material (adsorbent or resin) until all available active sites are saturated (e.g., Gkika et al., 2025). Thereafter, acids such as H_2SO_4 are added to remove metals and allow exchanger resin, zeolite, and bentonite to be reused to bind other metals (Mulopo, 2022). Despite ion exchange metal removal efficiency with metals such as Co, Cu, Zn, Ni, and Pb within higher pH (up to 11), it is however restricted to temperature, pH, and concentration of metals in AMD (Beauclair et al., 2024; Roy & Bhattacharya, 2015). Another restriction lies on the working pH range of 4–8 and the requirement to have low suspended solid and low concentration of Fe and Al in initial AMD (EPA, 2014).

4.1.1.3 Innovative and emerging solutions

With the advancement of technology, it is now possible to monitor AMD actively, that is, in real time. The active intervention means real-time monitoring through the use of controlled dosing systems (Naidu et al., 2021). This real-time monitoring of AMD is achieved through addition of controlled alkaline chemicals to adjust pH toward neutrality and precipitate out heavy metals (Naidu et al., 2021; Trumm, 2010). The systems involve use of sensors for pH detection, flowmeters to regulate flow rate, and metal concentration sensors to determine amount of heavy metals present in AMD to release the right ratio amount of alkaline chemicals for neutralization of pH and subsequently precipitate out metals in the form of insoluble compounds (Akcil & Koldas, 2006). Practically, the addition of chemicals (caustic soda, ammonia, lime, limestone) is done either after certain interval of time or batch by batch through utilization of pH and metal concentration sensors

Table 4.1 The chemical reagents used in AMD treatment.

No	Common name	Chemical name	Formula	2010 cost (US$/ton or gal)	2023 cost (US$/ton or gal)
1	Limestone	Calcium carbonate	$CaCO_3$	60/ton	100/ton
2	Hydrated lime	Calcium hydroxide	$Ca(OH)_2$	275/ton	200–300/ton
3	Pebble quicklime	Calcium oxide	CaO	355/ton	100–500/ton
4	Soda ash	Sodium carbonate	Na_2CO_3	840/ton	1000/ton
5	Caustic soda (solid)	Sodium hydroxide	NaOH	1240/ton	1140–1280/ton
6	Liquid caustic 20%	Sodium hydroxide	NaOH	0.95/gal	4.73/gal
7	Liquid caustic 50%	Sodium hydroxide	NaOH	2.35/gal	60/gal
8	Ammonia	Anhydrous ammonia	NH_3	1000/ton	1237/ton

AMD, acid mine drainage.
From Othman et al. (2023).

which in turn enables the release of required amount of alkaline chemicals that will mix with AMD to adjust pH toward neutrality and precipitate metals (Gaikwad & Gupta, 2007; Naidu et al., 2021). Some of the common alkaline chemical limes widely used in dosing for treatment of heavy metals, and that can adjust pH toward neutrality from the AMD are highlighted in Eqs. (4.9) and (4.10) (Table 4.1).

$$Fe^{3+} + 3OH^- \rightarrow Fe(OH)_2 \tag{4.9}$$

Additional hydrated lime consumes H^+, thereby neutralizing pH in the AMD:

$$2H^+ + Ca(OH)_2 \rightarrow Ca^{2+} + 2OH^- \tag{4.10}$$

Although active dosing is limited to high cost for building, operation, and maintenance (Othman et al., 2023; Trumm, 2010), it has demonstrated great efficiency of metal removal and neutralization of pH (Othman et al., 2023).

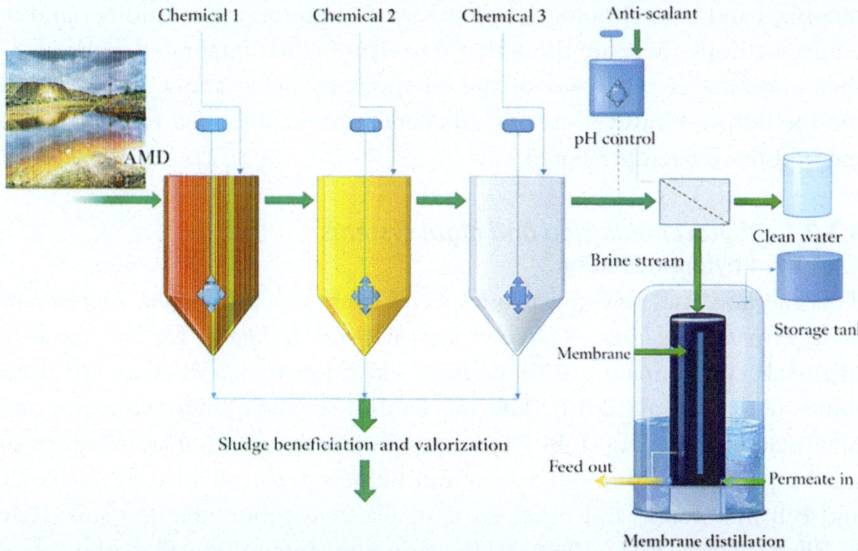

Figure 4.3 The use of chemicals in active treatments. Integration of active chemicals. *(From Nguegang and Ambushe (2024))*.

4.1.1.4 Integrated and interdisciplinary methods

Integrated and interdisciplinary methods entail active chemicals, active physical-chemical or physical treatment approaches (Beauclair & Ambushe, 2024)—they have demonstrated efficiency in removal of metals in AMD (Ryu et al., 2019) and pH neutralization (Beauclair & Ambushe, 2024; Mapukata et al., 2024). Fig. 3.6 demonstrates how integrated and interdisciplinary approaches treat AMD containing multiple heavy metals in multistages, whereby at each stage, certain chemicals are added to precipitate out specific heavy metal(s) with respect to pH required (Beauclair & Ambushe, 2024). For instance, integration of nanoparticle cryptocrystalline $MgCO_3$ and $BaCl_2$ where $MgCO_3$ neutralizes pH and $BaCl_2$ is removed as residual of sulfate by forming $BaSO_4$ (Mapukata et al., 2024; Masindi et al., 2015). As a general case, the integration approach of active chemicals with physical treatment methods (Fig. 4.3) has proven efficiency for neutralization of pH and removal of metals such as Al, Cu, Zn, at 99.7%, 98.6%, 98.2%, and 65.5% for Fe, Cu, Mn, and Cd, respectively (Beauclair & Ambushe, 2024).

4.1.2 Passive methods

In this subchapter, one may note that some of the methods are similar to those in the previous subchapter (i.e., on active methods), particularly microbial and biotechnology, innovative and engineering solutions—and in either methods, there are those that we term them as integrated methods—they combine one or two of the methods to bring about efficiency of the method(s). However, in this subchapter, more advanced forms of such methods have been presented.

4.1.2.1 Phytoremediation and algal systems

4.1.2.1.1 Phytoremediation

This method uses selected plant species such as *Chrysopogon, Limnocharis flava, Hyparrhenia hirta* , *Cynodon dactylon*, and *Eichhonia* for the removal of metals contaminating soils, tailings, and other materials extracted from mines (Ansari et al., 2016). The mechanism involves phytoextraction and phytostabilization (Fig. 4.4). Phytoextraction is the removal of metals from soil by plant roots through xylem and phloem into cell vacuole, cell wall, and cell membrane and other parts of plants without affecting soil (Roy & Bhattacharya, 2015). Phytostabilization concentrates metals within plant rhizosphere, thus removing metals from soil. This specie is termed as "metallophyte" since it accumulates metals (Burges et al., 2018).

Ansari et al. (2016) used Mediterranean plant species for treating Cd in soil produced from mine tailings. They used *Lygeum spartum, Atriplex halimus, Helichrysum stoechas, Dittrichia viscosa, Piptatherum miliaceum*, and *Limonium cossonianum* supported with pig slurry, pig manure, and marble waste. These species were planted to remove Cd by phytoextraction or phytoremediation. The result demonstrated improved in soil pH and lowing Cd in soil. Initially, the pH was 5.85, but after treatment, the pH raised to 7.71, while the initial concentration of Cd was 20.82 mg/kg, which was lowered to less than 3 mg/kg, which is below WHO toxicity limit (Ansari et al., 2016). Phytoremediation method is cost effective as it entirely depends on plant natural processes to remove metals from AMD. In other words, it does not rely on the addition of a substrate (Adetunji & Erasmus, 2025). But the phytoremediation efficiency varies from species to species of plant(s) and relies on the depth of the root of plant into the ground and soil properties (Ansari et al., 2016). Further, it is relatively slow since it depends onto the physiological process of the plants (Adetunji & Erasmus, 2025; Lee, 2023). Metal removal efficiency by various plant species is provided in Table 4.2.

4.1.2.1.2 Algal system

The algal system uses microalgae to adjust pH toward neutrality and remove metals from AMD through bio-absorption (Cai et al., 2021;

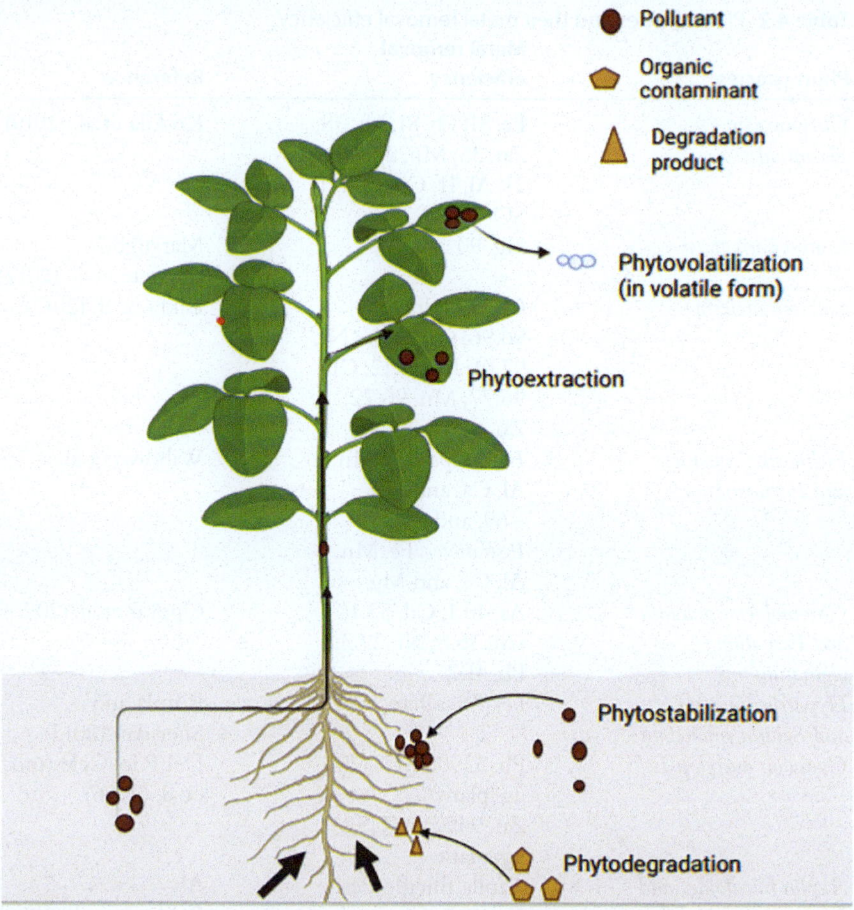

Figure 4.4 Phytoremediation. Schematic diagrams showing mechanisms of removal of pollutants from acid mine drainage by plants, that is, phytoremediation. *(From Adetunji and Erasmus (2025)).*

Dos Santos et al., 2020) or bioaccumulation (Bwapwa et al., 2017). Some of the microalgae species in this regard are *Spirulina, Chlorella, Scenedesmus, Cladophora, Oscillatoria, Anabaena,* and *Phaeodactylum*. When microalgae remove metals in AMD through contact with metals outside their cells by interacting between metal ions and anionic cell ligands, microprecipitation and covalent bonds and protein or polymer, is said to be removing metals through "extracellular mechanism" (Bwapwa et al., 2017; Du et al., 2022). Extracellular or intracellular mechanisms control the metals removal in microalgae systems. The cell walls of algae contain functional groups such as proteins, chelatin, and lipid which have affinity to metals (Fig. 4.5).

Table 4.2 Plant species and their metal removal efficiency.

Plant species	Metal removal efficiency	Reference
Chrysopogon zizanioides	Fe, 81; Pb, 81; Ni, 38; Zn, 35; Mn, 27; Cr, 21; Al, 11; Cu, 8; and $SO^{2-}{}_4$, 28	Kiiskila et al. (2019)
Limnocharis flava	Hg, 90	Marrugo-Negrete et al. (2017)
Salvinia molesta	Cd, 80.99; Pb, 96.96; Cr, 92.85; Ni, 97.01; Fe, 94.12; Cu, 96.77; Mn, 96.22; Zn, 96.38	Lakra et al. (2017)
Eichhornia crassipes and Pistia stratiotes	*E. crassipes*: Fe, Mn, Al, Ca, and Mg ~69; and *P. stratiotes*: Fe, Mn, Al, Ca, and Mg ~62	Wibowo et al.
Chromolaena ordorata and Paspalum viginatum	As, 40.1; Cd, 53.1; Cu, 36.8; Ni, 29.3; Pb, 41.2	Opoku et al. (2020)
Hyparrhenia hirta and Setaria sphacelata	Fe, 99; sulfate, 80	Ramla and Sheridan (2015)
Cynodon dactylon L.	Pb, 3290.8–15,520.1 μg/plant Zn, 9497.4–17,573 μg/plant	Del Río-Celestino et al. (2006)
Azolla filiculoides and Lemna minor	Azolla filiculoides: Cu, 100 Lemna minor: Cu, 74; Cr (VI), 99.5	Al-Baldawi et al. (2022)

From Adetunji and Erasmus (2025).

Unlike extracellular mechanism, intracellular mechanism is characterized by metal accumulation within vacuoles, cell walls, mitochondria, and chloroplast and then binds with proteins, lipids, or DNA of cells (Chen et al., 2011). The intracellular accumulation of metals within cells of microalgae is relatively very slow compared to extracellular accumulation and is very selective to some species of microalgae, such as *Scenedesmus* (Dos Santos et al., 2020). The metal accumulation within the cells of algae and therefore the efficiency of metal removal is influenced by the physiological conditions, such as pH and temperature of microalgae, metal type, and algae species

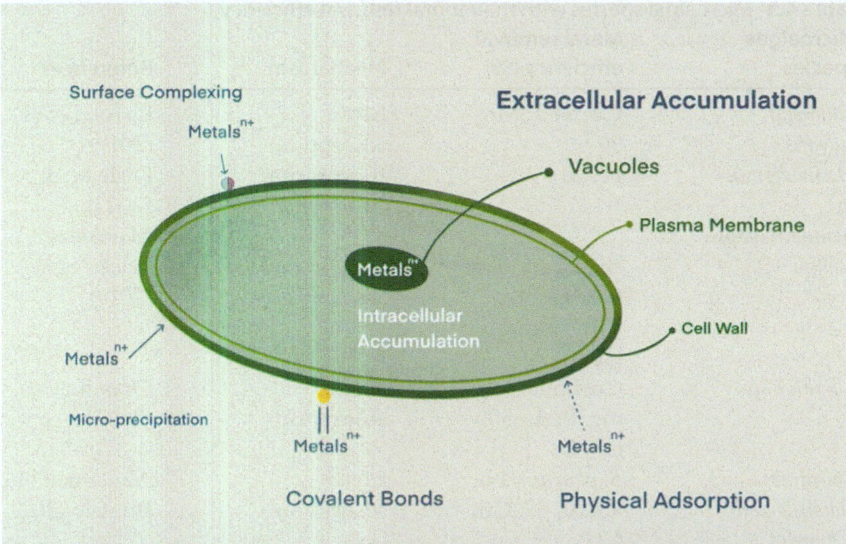

Figure 4.5 The microalgae and their power in AMD treatment. Extracellular mechanism by microalgae. *AMD*, Acid mine drainage. *(From Du et al. (2022)).*

(Monteiro et al., 2012). Likewise, high concentration of metal ions within microalgae varies proportional to rate of removal by algae (Monteiro et al., 2011) and concentration of algae increases with increase in rate of metal removal (Monteiro et al., 2011). The microalgae capability of metal removal from AMD for various algae species differs (Table 4.3). For instance, chlorophyte has higher capacity than Phaeophyta, and Phaeophyta have higher metal removal capability than *Rhodophyta* (Du et al., 2022). Aksu (2002) and Monteiro et al. (2012) have shown that some algae species such as *Chlorella vulgaris* work effectively in high temperature whereas *Oedogonium* and *Sargassum* species have low metal removal rate in high temperatures (Cruz, 2004).

In summary, using algae to treat AMD is a low-cost method because algae are easy to grow and widely available. However, the success of this method depends heavily on environmental conditions—especially pH, temperature, and the availability of nutrients (Adetunji & Erasmus, 2025; Du et al., 2022).

4.1.2.2 Microbial and biotechnological approaches

Passive microbial treatment relies on natural biological processes and ambient environmental conditions to achieve outcomes that are similar to those of phytoremediation (e.g., van Epps, 2006), algae, and other passive

Table 4.3 Microalgal species with their metal removal efficiency.

Microalgae species	Metal removal efficiency (%)	Mechanism	References
Chlorella vulgaris	Cu, 64.7; Mo, 99.9	Metal adsorption	Urrutia et al. (2019)
Desmodesmus sp.	Fe, 90	Biosorption	Diale et al. (2023)
Nannochloropsis oculata	Bioaccumulation: Cu, 89.29 Adsorption: Cu, 10.70	Bioaccumulation; adsorption	Martínez-macias et al. (2019)
Ulothrix sp.	Cu, 3500 mg/L; As, 500 mg/L	Metal absorption	Orandi et al. (2007)
Spirulina platensis and Chlorella vulgaris	*S. platensis*: Ni, 95; Al, 87; Cu, 62 *C. vulgaris*: Ni, 87; Al, 79.1; Cu, 80	Metal adsorption	Almomani and Bohsale (2021)
Spirulina sp.	Sulfate, 89; Fe, 99; Pb, 95; Zn, 93; Cu, 94	Alkalinity generation and metal precipitation	Van Hille et al. (1999)

From Du et al. (2022) and Adetunji and Erasmus (2025).

treatment systems with minimal external chemical and energy inputs (e.g., Sharma et al., 2024). In this approach, AMD is treated in systems such as constructed wetlands, compost bioreactors, or anaerobic cells that are designed to function with limited human intervention. The microorganisms, often introduced into organic-rich media such as compost, manure, or wood chips, operate under natural gradients of temperature and pH, utilizing slow-release carbon sources as electron donors. Although the biochemical mechanisms involved are fundamentally the same as those in active systems, passive microbial systems, in particular, operate more slowly and depend on the sustainability of microbial populations in a less controlled environment. In such a system, SRB reduce sulfate to hydrogen sulfide, precipitating metals as sulfides, while bicarbonate formation helps to neutralize acidity. Iron-oxidizing bacteria naturally present in the system also contribute to iron oxidation and proton consumption, enhancing overall pH improvement (e.g., Novair et al., 2024).

Species such as *Desulfosporosinus, Desulfitobacterium, Alicyclobacillaceae, Peptococcaceae*, and *Thermodesulfobium* have been identified in passive treatment environments as capable of thriving under suitable but unregulated conditions (Dev et al., 2021; Nobahar et al., 2024; Phyo et al., 2020; Sanchez-Andrea et al., 2015). Their activity however is strongly influenced by factors such as pH, ambient temperature, presence or absence of oxygen, pollutant type and the availability of organic substrates (Adetunji & Erasmus, 2025). While passive systems are generally more sustainable and cost-effective over long periods, their performance can be variable and often slower than that of active systems. Nevertheless, they remain a critical component of AMD remediation strategies, particularly in remote or low-resource settings where ongoing operational support is limited. Microbes especially bacteria are used in bioremediation (Table 4.4).

4.1.2.3 Innovative and emerging solutions

4.1.2.3.1 Self-sustaining, low-maintenance bioreactors, or natural treatment approaches

Developing a remediation that is less expensive with long sustainability operation is important for continuous treatment of AMD as compared to chemical-energy dependent techniques, which prove to be expensive (Zagury & Neculita, 2007). The low and self-sustaining system use biochar, which are organic matter produced through pyrolysis at different temperatures (Wang, 2020). The biochar is a synthesis and modified through pyrolysis by heating with microwave at a certain temperature. To enhance the efficiency, biochar may be modified by introducing $KMnO_4$, H_2O_2, and $K_2Cr_2O_7$ to increase surface area and reactivate functional group(s) to fix metals from AMD through adsorption (Shirvanimoghaddam, 2022; Yin, 2022). Furthermore, hydrothermal pyrolysis is used to treat feedstock at 120°C–250°C to form biochar with more functional groups to fix heavy metals from AMD (Wang, 2020). The widely used biochar is hardwood, corn straw, sugar cane, sewage sludge, paper mill sludge, etc. (Table 4.4)—these materials (biochar) can be outsourced easily in environment for use in treatment of AMD (Du et al., 2022).

The removal of heavy metals by biochar involves physical sorption, ion exchange, precipitation, complexation, and electrostatic processes (Du et al., 2022; Tan et al., 2015). The surface precipitation involves exchange of metal ions with biochar functional group anions, such as $CO_3{}^{2}$, $PO_4{}^{3-}$, and OH^- (Fig. 4.6). Cai et al. demonstrated the treatment of Cd by using orange peel through surface precipitation. Furthermore, Cu was removed by use

Table 4.4 Metal removal efficiencies by various biochar.

Biochar	Temperature for biochar production (°C)	AMD composition (mg/L)	Metal removal efficiency	References
Hardwood	450	Cu, 256; Zn, 260	Cu, 6.8 mg/g Zn, 4.5 mg/g	Chen et al. (2011)
Corn straw	600	Cu, 256; Zn, 260	Cu, 12.5 mg/g Zn, 11.0 mg/g	Chen et al. (2011)
Corn straw	400	Cd, 20; Pb, 20	Cd, 38.8 mg/g Pb, 29.0 mg/g	Chi et al. (2017)
Hickory wood	Pre-treated by $KMnO_4$ and then 600°C pyrolyzed	Pb, 100; Cu, 30; Cd, 30	Pb, 153 mg/g Cu 34.2 mg/g Cd 28.1 mg/g	Wang et al. (2015)
Sugar cane	500	Pb, 6.0–223	Pb, 87.0 mg/g	Abdelhafez and Li (2016)
Orange peel	500	Pb, 6.0–223	Pb, 6.0–223	Abdelhafez and Li (2016)
Almond shell	650	Ni and Co, 50–200	Ni, 22.2 mg/g Co, 28.1 mg/g	Kiliç et al. (2013)
Sewage sludge	550	Pb, 100–1000	Pb, 30.88 ± 0.95 mg/g	Lu et al. (2012)
Peanut straw, soybean straw, Canola straw	400	Cu, 15–960	Cu, 37.12–89.6 mg/g, peanut > soybean>canola	Tong et al. (2011)

(*continued on next page*)

White birch, black spruce	454, followed by KOH, CO_2 and steam activation	Cu, 100	Cu, >99%	Braghiroli et al. (2019)
Papermill sludge	270–720	As, 22.7; Cd, 33.0	As, 22.8 mg/g Cd, 41.6 mg/g	Yoon et al. (2017)
Nutshells, plum stones, wheat straws, grape stalks, and grape husks	600	Cd, 11.2–168 Pb, 20.7–310.5	Over 95% removal efficiency for all four biochar	Trakal et al. (2014)
Pistachio green hull	450	Cu, 70–270	Cu, 19.8 mg/g (62%)	Jalayeri and Pepe (2019)
Olive pomace	Hydrothermal 300, pyrolysis 300 and 600	Cu, 20	Cu, 77.8%	Pellera et al. (2012)
Lolium perenne, L. perenne fiber	300, 450, and 600 slow-pyrolysis process	Zn, 18.5 ± 2.10	*L. perenne* fiber has the best performance removal of Zn, 93.0%	Hodgson et al. (2016)
Platanus orientalis Linn leaves	400, modified by H_2O_2, $KMnO_4$, and $K_2Cr_2O_7$	Cd, 50	$KMnO_4$ modified biochar reached the highest removal efficacy of Cd, 54.7 mg/g	Yin et al.

(continued on next page)

Table 4.4 Metal removal efficiencies by various biochar—cont'd

Biochar	Temperature for biochar production (°C)	AMD composition (mg/L)	Metal removal efficiency	References
Rice straw and Fe_3O_4 and $CaCO_3$	400	Cd and As, 10–60	Cd, 6.34 mg/g As, 10.1 mg/g	Wu et al. (2018)
Rice husks	300, 500, and 700	Pb (concentration unknown)	Pb, RH300 14.1 mg/g, RH500 21.7 mg/g, RH700 26.7 mg/g	Shi et al. (2019)
Jarrah and pine wood chips	700	Cu, 17.3–195 Zn, 17.6–173	Cu, 4.39 mg/g Zn, 2.31 mg/g	Jiang et al.
Oakwood, Oakbark	400 and 450, followed by magnetic activation	Pb and Cd, 0–100	Pb, 100% Cd, 53%–99%	Mohan et al. (2014)
Orange peel	400, 500, 600, 700, 800	Cd, 100	Cd, 115 mg/g	Tran et al. (2016)

(continued on next page)

Sewage sludge	500 (for sewage sludge [Cs]), $ZnCl_2$ activated (for sludge-based active carbon [SBAC]), modified by nitric acid at different concentration and temperature (MSBACs)	Pb, 100–200	Pb, MSBAC 26.6 mg/g SBAC 17.0 mg/g CS, 4.42 mg/g	Li
Dairy manure	350	Cu, 0–320 Zn, 0–325 Cd, 0–560	Cu, 54.4 mg/g Zn, 32.8 mg/g Cd, 51.4 mg/g	Xu et al. (2013)
Poultry litter	400	Al, 51 Cu, 30.7 Zn, 26.8	Al, 100% Cu, 100% Zn, 99%	Oh and Yoon (2013)
Sesame straw	700	Pb, Cu, Zn and Cr are all 2.5–320	Pb, 102 mg/g Cu, 55.0 mg/g Zn, 34.0 mg/g Cr, 65 mg/g	Park et al. (2016)

(continued on next page)

Table 4.4 Metal removal efficiencies by various biochar—cont'd

Biochar	Temperature for biochar production (°C)	AMD composition (mg/L)	Metal removal efficiency	References
Rice husks	Hydrothermal 300, pyrolysis 300 and 600	Cu, 20.0	Cu, 90.1%	Pellera et al. (2012)
Common reed (Phragmites australis)	450	Fe, 0.36 and 28.8 Al, 0.13 and 10.99 Ni, 0.07 and 0.39 Zn, 0.03 and 0.19 Mn, 0.37 and 5.08	Metal removal by 89.0%–98.0% (Fe≈Al>Ni≈Zn>Mn)	Mosley et al. (2015)
Sludge	300 (nano zero-valent)	Sb, 10, 20, and 30	Sb, 160.40 mg/g	Wei et al. (2020)
Soy sauce residue	400 and modified by nanoscale FeS and chitosan	Cr, 100–550	Cr, 70.42 mg/g (76.07%)	Yang et al. (2021b)
Oakwood, Oak bark, Pinewood, Pine bark	400 and 600, fast pyrolysis	Cd, As, and Pb are all 0.01–0.10	Oak bark has the highest removal efficiency Pb, 11.4 mg/g	Mohan et al. (2007)

(continued on next page)

Aloe vera shell	700 followed by $Ni_{0.5}Zn_{0.5}Fe_2O_4$ magnetic nanoparticles supported	Ag, 100	Ag, 98.3% (244 mg/g)	Beigzadeh and Moeinpour (2016)
Spent coffee grounds	400	Cd, 0.228 Cu, 0.194 Pb, 0.0156 Zn, 0.0222	Cd, 99% Cu, 88% Pb, >99% Zn, 99%	Kim et al. (2014)
Canna indica	300, 400, 500 and 600	Cd, 30–200	Cd, 189 mg/g	Cui et al. (2016)
Coconut shell	400, $MgCl_2$ modification	Pb, 1000 Cd, 1000	Pb, 271.53 mg/g Cd, 91.95 mg/g	Wu et al. (2021)
Durian shell, *Robinia pseudoacacia*	500, Fe/Zn modification	Cd, 30–300	Cd, Durian shell biochar, 99.81% *R. pseudoacacia* biochar, 71.08%	Yang et al. (2021)

AMD, acid mine drainage.From Du et al. (2022).

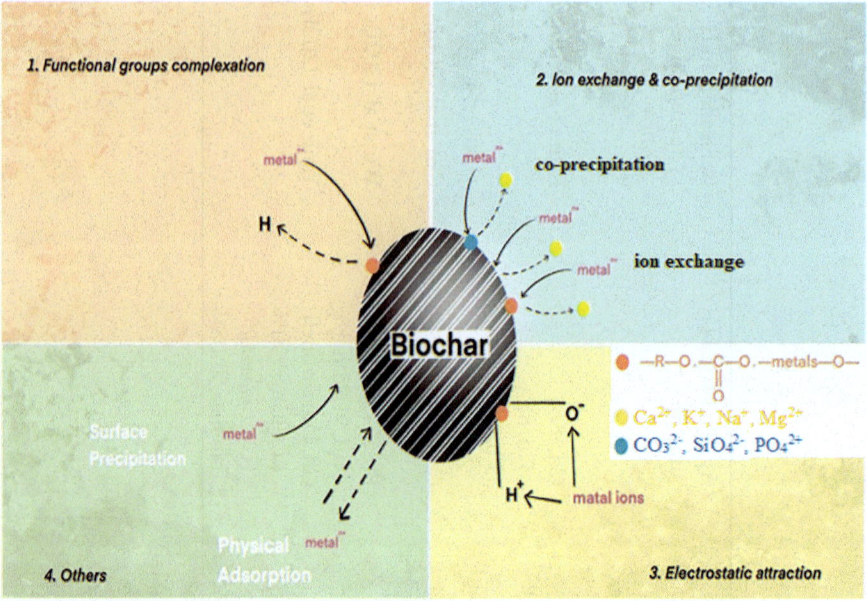

Figure 4.6 Mechanism of metal removal by biochar. *(From Tan et al. (2015) and Du et al. (2022)).*

of pistachio green hull biochar through surface precipitation (Jalayeri & Pepe, 2019). For ion exchange, electrostatic interaction and complexation are influenced by pH which influences interaction of biochar functional group (carboxyl, hydroxyl, or amino group) with heavy metals such as Mg, Ca, Pb, and Cr where optimal pH for metals removal is typically 6–7 with oxygenated functional group of C– –O, C–O, and phenolic O–H (Abdelhafez & Li, 2016). Thus, because of the complexation between oxygenated functional group and heavy metals, neutralization occurs by uptake of H^+ (Ding, 2016). Metals are then removed by biochar of pine wood trough complexation, where pH at around 1 favors the electrostatic interaction (Abdelhafez & Li, 2016). While metals such as Cu, Cd, Zn, Fe, Al, As, and Mn are removed by physical absorption between surface of biochar and heavy metals, the physical sorption process is not affected by pH (Oh & Yoon, 2013). The use of low-cost modified biochar with $KMnO_4$ or $K_2Cr_2O_7$ has demonstrated to have greater metal removal (Yin, 2022). Furthermore, the biochar works effectively even at low pH such as 2 to remove metals such as Pb (Yang, 2021b). For some metals such as Cu and Cd, they require optimum pH of 5–6 for them to be removed from AMD (Chen et al., 2011; Fig. 4.6).

4.1.2.3.2 Limitation

The biochar treatment of heavy metals from AMD is influenced by parameters such as pH, zero-point charge of biochar, and temperature.

4.1.2.3.3 Covers with capillary barrier effects

Covers with capillary barrier effects (CCBEs) are engineered, multilayered systems designed to isolate reactive waste (such as sulfidic tailings) from being in contact with the environment (Bussière et al., 2003; Rahardjo et al., 2019). The CCBEs are designed to limit downward infiltration of water and oxygen influx to tailings, thus preventing sulfide oxidation and leaching (Xiong et al., 2025). The CCBEs employ the difference in hydraulic properties between fine-grained and coarse-grained soil layers to keep a one-layer nearly saturated with water (Dublet-Adli et al., 2021). Capillary movement of water is the core principle of CCBEs (Dublet-Adli et al., 2021; Rahardjo et al., 2019). When two materials differing in their grain size are placed together in contact with water, water moves towards the fine particles irrespective of gravity force (Rahardjo et al., 2019). The CCBEs are made up of three or more layers (Aubertin et al., 1995). For example, in a six layered CCBE proposed by Aubertin et al. (1995), the layers are (from top to bottom) (1) surface layer, characterized by erosion protection (such as vegetation, topsoil, or riprap) and providing a buffer against weather and biointrusion. The surface layer is made up of fertile soil to support vegetation and it is designed to prevent wind and water erosion (Bussière et al., 2003); (2) protection layer, characterized by being composed of soil with moderate permeability that acts to store water for plants, shield the lower layers from freezing/thawing, and resist erosion, often designed to reduce evaporation. The protection layer is designed to be mechanically strong to offer protection against biointrusion, erosion, and effects of dry-wet cycles (Bussière et al., 2003). It is made up of coarse sand and gravel; (3) drainage layer, characterized by high hydraulic conductivity (coarse material like sand or gravel) to facilitate horizontal drainage and prevent water buildup on the underlying low-permeability layer; (4) moisture retaining layer (MRL) as fine-grained layer, characterized by low permeability and high water-holding capacity (clay or silt), ensuring it remains saturated or close to saturation, which prevents oxygen and water infiltration into the tailings. The MRL function is to prevent the influx of oxygen to the reactive mine waste. This is achieved by keeping the layer saturated with water (Aubertin et al., 1995). The degree of saturation of about 85%–90% is recommended (e.g., Kabambi et al., 2017). Oxygen moves through water roughly 10,000 times slower than it does through air

(Lacroix et al., 2022; Schlüter et al., 2025). The MRL layer must be med up of smaller particles than those of capillary break layer (CBL), usually fine grained material such as silty and clay soil Aubertin et al., 1995; (5) support layer, characterized by being a stable foundation that prevents settling or cracking of the above layers and provides a transition zone for the capillary barrier; and (6) tailings—the waste material being covered (Aubertin et al., 1995; Bussière et al., 2003). Although the CCBEs are made up of the six layers, only three layers are important for preventing AMD (Xiong et al., 2025) —these are (1) the MRL that lies between the two CBLs, (2) the one on top, and (3) the other at the bottom (Kabambi et al., 2017). The bottom CBL has three functions: (1) to offer mechanical support for MRL and (2) to prevent upward capillary rise of contaminated water from the mine wastes. This phenomenon is called capillary break (Aubertin et al., 1995; Bussière et al., 2003). The bottom CBL is therefore made up of large particles such as gravel and coarse sand to break capillary rise (Kabambi et al., 2017). The upper CBL functions as a drainage layer. It drains water laterally from the CCBEs systems, consequently reducing pore pressure and preventing desaturation of the MRL. The upper CBL is made of coarse material such as sand and gravel, is designed to be about 30–100 cm thick, and has a gentle slope. These properties ensure sufficient drainage.

4.1.2.3.4 Limitations of covers with capillary barrier effects

Despite their strength in effectively preventing the oxygen ingress to sulfidic tailings using natural geological materials such as sand gravel and fine sand, CCBEs have some limitations which are: (1) Pre-oxidized tailings often contain high amounts of Fe^{3+} and acidic microbial communities, which can catalyze sulfide oxidation even in anoxic (oxygen-depleted) environments, a process known as indirect oxidation; (2) Lack of sufficient material for constructing CCBEs in some area like Quebec, Canada where fine soil is limited can increase the reclamation cost and (3) CCBEs are sensitive to climate change.

4.1.3 Integrated and interdisciplinary methods

This approach combines two or more passive methods for the treatment of AMD (Mapukata et al., 2024). It integrates microorganism (bacteria) through constructed wetland. The bacteria use their metabolic activity to remove metals in AMD and adjust pH toward neutrality. Sometimes the wetland may be incorporated with limestone for efficiency. The treatment using SRB convert dissolved sulfates to H_2S which react with metals forming $MS_{(s)}$ that precipitate out and thereafter removed. This integration is efficient in removing metals like Al, Cu, Fe, As, Cd, Zn, and neutralizing pH (RoyChowdhury et al., 2015). Recent review work by Mafane et al. (2025) on active, semiactive or integrated approach to reduce AMD using SRB shows that SRB-based bioreactors can achieve up to 93.97% of sulfate reduction, with metal recovery rates of 95% for Ni, 98% for Fe and Cu, and 99% for Zn under optimized conditions. Furthermore, they show that bioreactors that used glycerol and ethanol as a carbon source improved the efficiency of sulfate reduction, achieving a pH neutralization from 2.8 to 7.5 within 14 days of hydraulic retention time. They conclude that despite promising results achieved so far, several challenges remain. These include the need for optimal environmental conditions, the management of toxic hydrogen sulfide production, and the economic feasibility of large-scale applications (Table 4.5).

In summary, nonconventional AMD treatment methods provide a diverse and increasingly sophisticated set of remediation options that extend well beyond traditional chemical neutralization. As summarized in Table 4.6, active systems such as microbial bioreactors, electrochemical processes, membrane filtration and nanotechnology-based treatments offer rapid reaction rates and high removal efficiencies for a wide range of metals and sulfates. However, these systems often require substantial energy input, operational control, and maintenance. In contrast, passive approaches including phytoremediation, algal systems, biochar-based sorption, and passive microbial processes rely on natural mechanisms and low external inputs, offering long-term sustainability and cost effectiveness, albeit at slower treatment rates and with performance that is more sensitive to site-specific conditions.

Integrated and interdisciplinary treatment systems bridge the gap between active and passive approaches by combining chemical, biological, and physical processes to improve treatment robustness, efficiency, and adaptability. The incorporation of support technologies such as artificial intelligence further enhances AMD management by enabling predictive monitoring, optimizing process control and supporting informed decision-making. Ultimately, the successful application of nonconventional AMD treatment methods depends on careful evaluation of site-specific factors including drainage chemistry, flow regime, climate, land availability and long-term closure objectives. Hybrid solutions, informed by predictive tools and supported by lifecycle cost considerations, represent a promising pathway toward sustainable and effective AMD remediation.

Table 4.5 The effectiveness of various SRB-based bioremediation technologies for the treatment of acid mine wastewater.

Treatment technology	Constant variable	Electron donor	Effectiveness of technology
Biological sulfate-reducing column	Temperature = 30°C, HRT = 14 days and pH = 5.5	Sodium lactate	Sulfate reduction of 79.04%; 64.78%; and 50.27% using chicken dairy manure and sawdust as organic substrates, respectively. 5% of sulfide precipitation and over 95% of heavy metals.
Sulfidogenic fixed-bed column bioreactor	Temperature = 30°C, HRT = 7 days and pH = 4.5	Ethanol	Recovery of Sb at 97.8%; As 98.2%; and Fe(ii) was recovered at 85%.
Sulfate-reducing bacteria Cu/Fe reactor	Temperature = 37°C–45°C; pH = 1, HRT = 48 h COD rate of 27.4 mg COD/(Lh) and nitrate rate of 17.4 mg N/(Lh)	Sodium lactate	High heavy metal removal efficiency such as Cu^{2+} at 98.17; Zn^{2+} 99.67; and Cu at 7.5%–10% pH increase rate = 5–9.
Biochemical passive bioreactor	Temperature = 35°C, pH = 4, and HRT = 2 days	Mushroom compost, limestone, and cow manure	Sulfate recovery of greater than 60% and metal ions such as $Fe^{2+}$and Zn^{2+} of around 95%.
Sulfate-reducing wetland bioreactors	pH = 6.5 HRT of 6 days Lignocellulosic waste as a substrate	Lactate	Sulfate reduction of 60.7%, with COD removal of 70.6% and recovery of valuable metals such as Fe 99.6%; Zn 99.4%; Mn 9.3% Cd 99.9% and Cu 94.5%. pH increase rate = 6.5–7.7.

(continued on next page)

Downflow structure bed bioreactor	The temperature was kept at 30°C, pH = 4, and HRT = 19 h	Sugar (sugarcane)	Sulfate removal rate of 55–91%, 80% removal for Co, Ni and Zn; Cu 73%, Fe 70%, and Mn at 60%. pH increase rate = 6.7–7.5.
Fluidized bed bioreactor	Temperature = 20°C, pH = 7–8, HRT = 10 h, and COD/sulfate rate = 2.5 − 1.7	Glycerol	Sulfate removal rate of 80%–92%, COD removal rate of 58%, and recovery of metals such as Ni, Mn, and Cu at a rate of 90%.
Sulfate-reducing bioreactor	Temperature = 30°C, pH = 5, and HRT = 6 h	Iron sulfate	Metal iron recovery such as Zn(ii) 88.5%; Pb(ii) 92.6% Cu(ii) 76.0%; Mn(ii) 62.2%; Fe(iii) 56.9; Cd(ii) 78.7% and Ni(ii) 62.5. pH increase rate = 6.5.

SRB, sulfate-reducing bacteria; COD, Chemical Oxygen Demand.Modified from Mafane et al. (2025).

Table 4.6 Comparative summary of nonconventional AMD treatment.

Treatment method	Classification	Principal removal/neutralization mechanisms	Typical metals/parameters treated	Reported performance (from literature in this chapter)	Key advantages	Key limitations	Typical application scale
Microbial bioreactors (SRB-based)	Active (biological)	Sulfate reduction to sulfide; metal sulfide precipitation; alkalinity generation (HCO_3^-)	Fe, Cu, Zn, Mn, Cr, sulfate	Metal removal: Fe up to ~92%–93%, Cu ~80%–99%, Mn ~53%; pH increases from ~2.7 to ~6.2	High metal selectivity; simultaneous pH neutralization; potential metal recovery	Requires strict control of pH, temperature, redox; organic substrate supply; higher O&M costs	Bench, pilot and full-scale
Nanotechnology (engineered nanoparticles)	Active (physico-chemical)	Adsorption, ion exchange, surface complexation, nanofiltration	Cu, Cr, Ni, Zn, Cd, Pb, sulfate	Metal removal often >70–99% under optimal pH	Very high surface area; rapid kinetics; compatible with hybrid systems	Sensitive to pH and temperature; material cost; mostly lab-scale evidence	Laboratory and pilot
Electrochemical treatment	Active (electro-chemical)	Electrocoagulation, electroflotation, electrodeposition; cathodic H^+ consumption	Cu, Ni, Fe, Zn, Cr	Metal removal ~40%–99% depending on metal and operating conditions	No chemical dosing; low sludge; possible metal recovery	High energy consumption; electrode maintenance	Pilot and full-scale
Membrane filtration (NF, RO, UF, MF)	Active (physical separation)	Size and charge exclusion driven by pressure gradients	Dissolved metals, sulfate, TDS	NF ~93–98%; RO ~99% removal efficiency	Very high effluent quality; removal of organics and inorganics	Membrane fouling; concentrate disposal; capital cost	Pilot and full-scale

(continued on next page)

Ion exchange	Active (chemical)	Cationic, anionic or chelating resin exchange	Cu, Zn, Ni, Co, Pb, As, Fe, Ca, Mg	High removal efficiency at pH ~4–8	High selectivity; regenerable media	Sensitive to high Fe, Al and suspended solids; regeneration requires acids	Pilot and full-scale
Integrated active systems	Hybrid (chemical–physical–biological)	Sequential neutralization, filtration, adsorption, precipitation	Al, Fe, Cu, Zn, Mn, Cd	Removal often >95% for Al, Fe, Cu, Zn; Mn and Cd lower but significant	Synergistic performance; higher robustness	Increased system complexity; higher capital cost	Pilot and full-scale
Phytoremediation	Passive (biological)	Phytoextraction and phytostabilization	Fe, Pb, Cd, Zn, Mn, Cu, As	Metal removal from ~30% to >95% depending on species	Very low cost; environmentally sustainable	Slow kinetics; depth-limited roots; species-specific	Field, long-term
Algal systems	Passive (biological)	Biosorption, bioaccumulation, micro-precipitation	Cu, Ni, Zn, Fe, Pb, sulfate	Removal efficiencies ~60%–99% for selected metals	Low cost; easy cultivation; CO_2 uptake	Sensitive to pH, temperature, nutrients; harvesting required	Lab and pilot

(*continued on next page*)

Table 4.6 Comparative summary of nonconventional AMD treatment—cont'd

Treatment method	Classification	Principal removal/ neutralization mechanisms	Typical metals/parameters treated	Reported performance (from literature in this chapter)	Key advantages	Key limitations	Typical application scale
Passive microbial systems (wetlands, compost bioreactors)	Passive (biological)	Sulfate reduction, metal sulfide formation, buffering	Fe, Al, Zn, Cu, Mn, As, sulfate	Effective long-term metal attenuation and pH improvement	Low maintenance; sustainable; suitable for remote sites	Slower kinetics; performance variability	Field / full-scale
Biochar-based systems	Passive (sorption-based)	Physical sorption, ion exchange, surface precipitation, complexation	Cu, Cd, Pb, Zn, Ni, As, Mn	Often >80%–99% metal removal depending on feedstock and pH	Low-cost feedstocks; adaptable; effective even at low pH	Performance depends on pH, biochar type and modification	Lab and pilot
Integrated passive systems	Passive hybrid	Wetlands + SRB + limestone or organic media	Al, Fe, Cu, Zn, Cd, As	High long-term removal efficiencies (>90% for many metals)	Sustainable; minimal energy input	Requires land area; slower response	Field/full-scale

AI disclosure

During the preparation of this work, the author(s) used AI (CHAT GPT) to generate the Abstract of Chapters 4 and to correct English language in some parts of the text. After that, editing was manually done to ensure technical and professional elements were not compromised. The authors take full responsibility for the content of the publication.

References

Abdelhafez, A. A., & Li, J. (2016). Removal of Pb(II) from aqueous solution by using biochars derived from sugar cane bagasse and orange peel. *Journal of the Taiwan Institute of Chemical Engineers, 61*, 367–375. https://doi.org/10.1016/j.jtice.2016.01.005.

Abdullah, N., Yusof, N., Lau, W. J., Jaafar, J., & Ismail, A. F. (2019). Recent trends of heavy metal removal from water/wastewater by membrane technologies. *Journal of Industrial and Engineering Chemistry, 76*, 17–38. https://doi.org/10.1016/j.jiec.2019.03.029.

Abhanga, R. M., Wanib, K. S., Patil, V. S., Pangarkara, B. L., & Parjanea, S. B. (2013). Nanofiltration for recovery of heavy metal ions from waste water – A review. *International Journal of Research in Environmental Science and Technology, 3*(1), 29–34.

Adetunji, A. I., & Erasmus, M. (2025). Biological treatment of acid mine drainage: An eco-sustainable strategy for removal of toxic pollutants. *Journal of Hazardous Materials Advances, 18*. https://doi.org/10.1016/j.hazadv.2025.100659.

Agboola, O. (2019). The role of membrane technology in acid mine water treatment: A review. *Korean Journal of Chemical Engineering, 36*(9), 1389–1400. https://doi.org/10.1007/s11814-019-0302-2.

Akcil, A., & Koldas, S. (2006). Acid Mine Drainage (AMD): Causes, treatment and case studies. *Journal of Cleaner Production, 14*(12–13), 1139–1145. Issues https://doi.org/10.1016/j.jclepro.2004.09.006 .

Aksu, Z. (2002). Determination of the equilibrium, kinetic and thermodynamic parameters of the batch biosorption of nickel(II) ions onto Chlorella vulgaris. *Process Biochemistry, 38*(1), 89–99. https://doi.org/10.1016/S0032-9592(02)00051-1.

Al-Baldawi, I. A., Yasin, S. R., Jasim, S. S., Abdullah, S. R. S., Almansoory, A. F., & Ismail, N. I. (2022). Removal of copper by Azolla filiculoides and Lemna minor: Phytoremediation potential, adsorption kinetics and isotherms. *Heliyon, 8*(11). https://doi.org/10.1016/j.heliyon.2022.e11456.

Almomani, F., & Bohsale, R. R. (2021). Bio-sorption of toxic metals from industrial wastewater by algae strains Spirulina platensis and Chlorella vulgaris: Application of isotherm, kinetic models and process optimization. *Science of the Total Environment, 755*. https://doi.org/10.1016/j.scitotenv.2020.142654.

Ambiado, K., Bustos, C., Schwarz, A., & Bórquez, R. (2017). Membrane technology applied to acid mine drainage from copper mining. *Water Science and Technology, 75*(3), 705–715. https://doi.org/10.2166/wst.2016.556.

Ansari, A. A., Singh, S., Gill, G. R., Lanza, G. R., & Newman, R. (2016). *Phytoremediation management of environmental contaminants*: 4. Springer International Publishing. doi:10.1007/978-3-319-41811-7.

Bai, H., Kang, Y., Quan, H., Han, Y., Sun, J., & Feng, Y. (2013). Treatment of acid mine drainage by sulfate reducing bacteria with iron in bench scale runs. *Bioresource Technology, 128*, 818–822. https://doi.org/10.1016/j.biortech.2012.10.070.

Beauclair, N., Masindi, V., Msagati, T. A. M., & Memory, T. (2024). Acid mine drainage treatment technologies: Challenges and future perspectives. In *Hybridized technologies for the treatment of mining effluents* (pp. 245–285). Wiley. https://doi.org/10.1002/9781119896920.ch8.

Beigzadeh, P., & Moeinpour, F. (2016). Fast and efficient removal of silver (I) from aqueous solutions using aloe vera shell ash supported Ni0.5Zn0.5Fe2O4 magnetic nanoparticles. *Transactions of Nonferrous Metals Society of China (English Edition), 26*(8), 2238–2246. https://doi.org/10.1016/S1003-6326(16)64341-8.

Bejan, D., & Bunce, N. J. (2015). Acid mine drainage: Electrochemical approaches to prevention and remediation of acidity and toxic metals. *Journal of Applied Electrochemistry, 45*(12), 1239–1254. https://doi.org/10.1007/s10800-015-0884-2.

Boyd, G. R., Tuccillo, M. E, Sandvig, A., Pelaez, M., Han, C., & Dionysiou, D. D. (2013). Nanomaterials: Removal processes and beneficial applications in treatment. *Journal of American Water Works Association*. http://dx.doi.org/10.5942/jawwa.2013.105.0154.

Braghiroli, F. L., Bouafif, H., Neculita, C. M., & Koubaa, A. (2019). Performance of physically and chemically activated biochars in copper removal from contaminated mine effluents. *Water, Air, and Soil Pollution*, (8), 230. https://doi.org/10.1007/s11270-019-4233-7.

Brewster, E. T., Freguia, S., Edraki, M., Berry, L., & Ledezma, P. (2020). Staged electrochemical treatment guided by modelling allows for targeted recovery of metals and rare earth elements from acid mine drainage. *Journal of Environmental Management*, 275. https://doi.org/10.1016/j.jenvman.2020.111266.

Burges, A., Alkorta, I., Epelde, L., & Garbisu, C. (2018). From phytoremediation of soil contaminants to phytomanagement of ecosystem services in metal contaminated sites. *International Journal of Phytoremediation, 20*(4), 384–397. https://doi.org/10.1080/15226514.2017.1365340.

Bussière, B., Aubertin, M., & Chapuis, R. P. (2003). The behavior of inclined covers used as oxygen barriers. *Canadian Geotechnical Journal, 40*(3), 512–535. https://doi.org/10.1139/t03-001.

Bwapwa, J. K., Jaiyeola, A. T., & Chetty, R. (2017). Bioremediation of acid mine drainage using algae strains: A review. *South African Journal of Chemical Engineering, 24*, 62–70. https://doi.org/10.1016/j.sajce.2017.06.005.

Cai, T., Liu, X., Zhang, J., Tie, B., Lei, M., Wei, X., Peng, O., & Du, H. (2021). Silicatemodified oiltea camellia shell-derived biochar: A novel and cost-effective sorbent for cadmium removal. *Journal of Cleaner Production, 281*, 125390. https://doi.org/10.1016/j.jclepro.2020.125390.

Chartrand, M. M. G., & Bunce, N. J. (2003). Electrochemical remediation of acid mine drainage. *Journal of Applied Electrochemistry, 33*, 259–264.

Chen, X., Chen, G., Chen, L., Chen, Y., Lehmann, J., McBride, M. B., & Hay, A. G. (2011). Adsorption of copper and zinc by biochars produced from pyrolysis of hardwood and corn straw in aqueous solution. *Bioresource Technology, 102*(19), 8877–8884. https://doi.org/10.1016/j.biortech.2011.06.078.

Chi, T., Zuo, J., & Liu, F. (2017). Performance and mechanism for cadmium and lead adsorption from water and soil by corn straw biochar. *Frontiers of Environmental Science and Engineering, 11*(2). https://doi.org/10.1007/s11783-017-0921-y.

Cruz, C. C. V., da Costa, A. C. A., Henriques, C. A., & Luna, A. S. (2004). Kinetic modeling and equilibrium studies during cadmium biosorption by dead Sargassum sp. biomass. *Bioresource Technology, 91*, 249–257. https://doi.org/10.1016/S0960-8524(03)00194-9.

Cui, X., Fang, S., Yao, Y., Li, T., Ni, Q., Yang, X., & He, Z. (2016). Potential mechanisms of cadmium removal from aqueous solution by Canna indica derived biochar. *Science of the Total Environment, 562*, 517–525. https://doi.org/10.1016/j.scitotenv.2016.03.248.

Del Prado-Audelo, M. L., García Kerdan, I., Escutia-Guadarrama, L., Reyna-González, J. M., Magaña, J. J., & Leyva-Gómez, G. (2021). Nanoremediation: Nanomaterials and nanotechnologies for environmental cleanup. *Frontiers in Environmental Science, 9*. https://doi.org/10.3389/fenvs.2021.793765.

Del Río-Celestino, M., Font, R., Moreno-Rojas, R., & De Haro-Bailón, A. (2006). Uptake of lead and zinc by wild plants growing on contaminated soils. *Industrial Crops and Products, 24*(3), 230–237. https://doi.org/10.1016/j.indcrop.2006.06.013.

Dev, S., Galey, M., Chun, C. L., Novotny, C., Ghosh, T., & Aggarwal. , S. (2021). Enrichment of psychrophilic and acidophilic sulfate-reducing bacterial consortia—A solution toward acid mine drainage treatment in cold regions. *Environmental Science: Processes & Impacts, 23*, 2007–2020. https://doi.org/10.1039/D1EM00256B.

Diale, P. P., Hildebrandt, D., Glasser, D., Matambo, T. S., & Makgato, S. S. (2023). An analysis of the processes, kinetics and equilibrium of iron's biosorption on immobilized green microalgae. *South African Journal of Chemical Engineering, 45*, 210–220.

Aubertin, M., Chapuis, R.P., Mostafa, A., Bussiere, B., & Tremblay, J.-F.R. et al. (1995). Evaluation en laboratoire de barrieres seches construites a partir de residus miniers centre Canadien. Technical Report n° 1622. https://publications.polymtl.ca/32630/

Ding, Y., Liu, Y., Liu, S., Li, Z., Tan, X., Huang, X., Zeng, G., Zhou, Y., Zheng, B., & Cai, X. (2016). Competitive removal of Cd(II) and Pb(II) by biochars produced from water hyacinths: Performance and mechanism. *RSC Advances, 6*, 5223–5232. https://doi.org/10.1039/C5RA26248H.

Dlamini, C. L, De Kock, L. -A., Kefeni, K. K, Mamba, B. B, & Msagati, T. AM. (2019). Acid mine drainage pollution remediation using hybrid chelating ion-exchange/HZrO2 nanocomposite adsorbents. *SN Applied Sciences, 1*, 1618. https://doi.org/10.1007/s42452-019-1580-3.

Dlamini, N. G, Kotze Basson, A., & Pullabhotla, V. S. R. (2021). . Synthesis and application of FeCu bimetallic nanoparticles in coal mine wastewater treatment. *Minerals, 11*, 132. https://doi.org/10.3390/min11020132.

Dong, Y., Gao, Z., Di, J., Wang, D., Yang, Z., Wang, Y., & Xie, Z. (2023). Study on the effectiveness of sulfate reducing bacteria to remove heavy metals (Fe, Mn, Cu, Cr) in acid mine drainage. *Sustainability (Switzerland)*, (6), 15. https://doi.org/10.3390/su15065486.

Dorfner, K. (1991). *Ion-exchangers*. Walter de Gruyter.

dos Santos, K. B., de Almeida, V. O., Weiler, J., & Schneider, I. A. H. (2020). Removal of pollutants from an AMD from a coal mine by neutralization/precipitation followed by "in vivo" biosorption step with the microalgae Scenedesmus sp. *Minerals, 10*(8), 1–11. https://doi.org/10.3390/min10080711.

Du, T., Bogush, A., Mašek, O., Purton, S., & Campos, L. C. (2022). Algae, biochar and bacteria for acid mine drainage (AMD) remediation: A review. *Chemosphere, 304*. https://doi.org/10.1016/j.chemosphere.2022.135284.

Dublet-Adli, G., Pabst, T., Okkenhaug, G., Sætre, C., Vårheim, A. M., Tvedten, M. K., Gelena, S. K., Smebye, A. B., Kvennås, M., & Breedveld, G. D. (2021). Valorisation of partially oxidized tailings in a cover system to reclaim an old acid generating mine site. *Minerals, 11*(9), 1–20. https://doi.org/10.3390/min11090987.

Eniola, J. O., Kujawa, J., Nwokoye, A., Al-Gharabli, S., Avornyo, A. K., Giwa, A., Amusa, H. K., Yusuf, A. O., & Okolie, J. A. (2024). Advances in electrochemical membranes for water treatment: A comprehensive review. *Desalination and Water Treatment, 319*, 100450. https://doi.org/10.1016/j.dwt.2024.100450.

Dinardo, O., Kondos, P. D., Mackinnon, D. J., Mccready, R. G. L., Riveros, P. A., & Skaff, M. *Study on metals recovery/recycling from acid mine drainage phase IA: Literature survey*. http://mend-nedem.org/wp-content/uploads/2013/01/3.21.1a.pdf.

EPA. (2014). Reference guide to treatment technologies for mining-influenced water.Retrieved May 6, 2020, from https://nepis.epa.gov/Exe/ZyPDF.cgi/P100I4PB.PDF?Dockey=P100I4PB.pdf.

Feng, D., Aldrich, C., & Tan, H. (2000). Treatment of acid mine water by use of heavy metal precipitation and ion exchange. *Minerals Engineering, 13*(6), 623–642. https://doi.org/10.1016/S0892-6875(00)00045-5.

Gaikwad, R. W., & Gupta, V. (2007). *Acid mine drainage (AMD) management. Journal of Industrial Pollution Control, 23*(2), 285–297.

Gaikwad, R. W., Misal, S. A., & Gupta, D. V. (2009). Removal of copper ions from acid mine drainage (AMD) by ion exchange resins: Indion 820 and Indion 850. *Journal of Applied Sciences in Environmental Sanitation, 4*(2), 133–140.

Gkika, D. A., Tolkou, A. K., Katsoyiannis, I. A., & Kyzas, G. Z. (2025). The adsorption-desorption-regeneration pathway to a circular economy: The role of waste-derived adsorbents on chromium removal. *Separation and Purification Technology, 368*, 132996. https://doi.org/10.1016/j.seppur.2025.132996.

Gyu, B., Zhan, M., Shin, K., Lee, S., & Hong, S. (2017). Pilot-scale evaluation of FO-RO osmotic dilution process for treating wastewater from coal-fired power plant integrated with seawater desalination. *Journal of Membrane Science, 540*, 78–87. https://doi.org/10.1016/j.memsci.2017.06.036.

Hodgson, E., Lewys-James, A., Rao Ravella, S., Thomas-Jones, S., Perkins, W., & Gallagher, J. (2016). Optimisation of slow-pyrolysis process conditions to maximise char yield and heavy metal adsorption of biochar produced from different feedstocks. *Bioresource Technology, 214*, 574–581. https://doi.org/10.1016/j.biortech.2016.05.009.

Hu, J., Chen, G., & Lo, I. M. C. (2006). Selective removal of heavy metals from industrial wastewater using maghemite nanoparticle: Performance and mechanisms. *Journal of Environmental Engineering, 132*(7), 709–715.

Jalayeri, H., & Pepe, F. (2019). Novel and high-performance biochar derived from pistachio green hull biomass: Production, characterization, and application to Cu(II) removal from aqueous solutions. *Ecotoxicology and Environmental Safety, 168*, 64–71. https://doi.org/10.1016/j.ecoenv.2018.10.058.

Kabambi, K. A., Bussière, B., & Demers, I. (2017). Hydrogeological behaviour of covers with capillary barrier effects made of mining materials. *Geotechnical and Geological Engineering, 35*(3), 1199–1220. https://doi.org/10.1007/s10706-017-0174-3.

Kapoor, R. T., Salvadori, M. R., Rafatullah, M., Siddiqui, M. R., Khan, M. A., & Alshareef, S. A. (2021). Exploration of microbial factories for synthesis of nanoparticles – A sustainable approach for bioremediation of environmental contaminants. *Frontiers in Microbiology, 12*. https://doi.org/10.3389/fmicb.2021.658294.

Kefeni, K. K., Msagati, T. A. M., & Mamba, B. B. (2017). Acid mine drainage: Prevention, treatment options, and resource recovery: A review. *J. Clean. Prod., 151*, 475–493.

Khelifa, A., Moulay, S., & Naceur, A. W. (2005). Treatment of metal finishing effluents by the electroflotation technique. *Desalination, 181*, 27–33.

Kiiskila, J. D., Sarkar, D., Panja, S., Sahi, S. V., & Datta, R. (2019). Remediation of acid mine drainage-impacted water by vetiver grass (Chrysopogon zizanioides): A multiscale long-term study. *Ecological Engineering, 129*, 97–108. https://doi.org/10.1016/j.ecoleng.2019.01.018.

Kiliç, M., Kirbiyik, Ç., Çepelioğullar, Ö., & Pütün, A. E. (2013). Adsorption of heavy metal ions from aqueous solutions by bio-char, a by-product of pyrolysis. *Applied Surface Science, 283*, 856–862. https://doi.org/10.1016/j.apsusc.2013.07.033.

Kim, M. S., Min, H. G., Koo, N., Park, J., Lee, S. H., Bak, G. I., & Kim, J. G. (2014). The effectiveness of spent coffee grounds and its biochar on the amelioration of heavy metals-contaminated water and soil using chemical and biological assessments. *Journal of Environmental Management, 146*, 124–130. https://doi.org/10.1016/j.jenvman.2014.07.001.

Lacroix, E. M., Mendillo, J., Gomes, A., Dekas, A., & Fendorf, S. (2022). Contributions of anoxic microsites to soil carbon protection across soil textures. *Geoderma, 425*, 116050. https://doi.org/10.1016/j.geoderma.2022.116050.

Lakra, K. C., Lal, B., & Banerjee, T. K. (2017). Decontamination of coal mine effluent generated at the Rajrappa coal mine using phytoremediation technology. *International Journal of Phytoremediation, 19*(6), 530–536. https://doi.org/10.1080/15226514.2016.1267698.

Lau, W.-J., Ismail, A. F., Isloor, A., & Al-Ahmed, A. (2019). *Advanced nanomaterials for membrane synthesis and its applications* (p. 321). Elsevier. https://doi.org/10.1016/B978-0-12-814503-6.09993-6.

Lee, S.-H., Park, H., & Kim, J.-G. (2023). Current status of and challenges for phytoremediation as a sustainable environmental management plan for abandoned mine areas in Korea. *Sustainability, 15*, 2761.

Li, W.-W., & Yu, H.-Q. (2014). Insight into the roles of microbial extracellular polymer substances in metal biosorption. *Bioresource Technology, 160*, 15–23. https://doi.org/10.1016/j.biortech.2013.11.074.

Loshchinina, E. A., Vetchinkina, E. P., & Kupryashina, M. A. (2023). Diversity of biogenic nanoparticles obtained by the fungi-mediated synthesis: A review. *Biomimetics, 8*, 1.

Lu, H., Zhang, W., Yang, Y., Huang, X., Wang, S., & Qiu, R. (2012). Relative distribution of Pb2+ sorption mechanisms by sludge-derived biochar. *Water Research, 46*, 854–862. https://doi.org/10.1016/j.watres.2011.11.058.

Mafane, D., Ngulube, T., & Mphahlele-Makgwane, M. M. (2025). Anaerobic bioremediation of acid mine drainage using sulphate-reducing bacteria: Current status, challenges, and future directions. *Sustainability 2025, 17*, 3567. https://doi.org/10.3390/su17083567.

Mao, Q., Qin, W., Yan, B., & Luo, L. (2024). Ferrous iron oxidation efficiency and kinetics by indigenous iron-oxidizing bacteria in acid mine drainage. *Environmental Technology and Innovation, 36*. https://doi.org/10.1016/j.eti.2024.103785.

Mapukata, S., Mudzanani, K., Maurice Chauke, N., Maiga, D., Phadi, T., & Raphulu, M. (2024). Acid Mine drainage treatment and control: Remediation methodologies, mineral beneficiation and water reclamation strategies. *Hydrology—current research and future directions*. IntechOpen https://doi.org/10.5772/intechopen.1003848.

Mardonova, M., Shahid, M. K., Abbassi, R., Lim, J. W., Paudel, S. R., & Mainali, B. (2025). Chapter 2. Advances in acid mine drainage management through artificial intelligence. In *Artificial intelligence in future mining* (pp. 77–177). Elsevier. https://doi.org/10.1016/B978-0-443-28911-8.00002-0.

Marrugo-Negrete, J., Enamorado-Montes, G., Durango-Hernández, J., Pinedo-Hernández, J., & Díez, S. (2017). Removal of mercury from gold mine effluents using Limnocharis flava in constructed wetlands. *Chemosphere, 167*, 188–192. https://doi.org/10.1016/j.chemosphere.2016.09.130.

Martínez-Macias, M. D. R., Correa-Murrieta, M. A., Villegas-Peralta, Y., Dévora-Isiordia, G. E., Alvarez-Sánchez, J., Saldivar-Cabrales, J., & Sanchez-Duarte, R. G. (2019). Uptake of copper from acid mine drainage by the microalgae Nannochloropsis oculata. *Environmental Science and Pollution Research., 26*, 6311–6318.

Masindi, V., Foteinis, S., Renforth, P., Ndiritu, J., Maree, J. P., Tekere, M., & Chatzisymeon, E. (2022). Challenges and avenues for acid mine drainage treatment, beneficiation, and valorisation in circular economy: A review. *Ecological Engineering, 183*, 106740. https://doi.org/10.1016/j.ecoleng.2022.106740.

Masindi, V., Gitari, M. W., Tutu, H., & De Beer, M. (2015). Passive remediation of acid mine drainage using cryptocrystalline magnesite: A batch experimental and geochemical modelling approach. *Water SA, 41*(5), 677–682. https://doi.org/10.4314/wsa.v41i5.10.

Mohan, D., Kumar, H., Sarswat, A., Alexandre-Franco, M., & Pittman, C. U. (2014). Cadmium and lead remediation using magnetic oak wood and oak bark fast pyrolysis bio-chars. *Chemical Engineering Journal, 236*, 513–528. https://doi.org/10.1016/j.cej.2013.09.057.

Mohan, D., Pittman, C. U., Bricka, M., Smith, F., Yancey, B., Mohammad, J., Steele, P. H., Alexandre-Franco, M. F., Gómez-Serrano, V., & Gong, H. (2007). Sorption of arsenic, cadmium, and lead by chars produced from fast pyrolysis of wood and bark during bio-oil production. *Journal of Colloid and Interface Science, 310*(1), 57–73. https://doi.org/10.1016/j.jcis.2007.01.020.

Monteiro, C. M., Castro, P. M. L., & Malcata, F. X. (2011). Biosorption of zinc ions from aqueous solution by the microalga Scenedesmus obliquus. *Environmental Chemistry Letters, 9*(2), 169–176. https://doi.org/10.1007/s10311-009-0258-2.

Monteiro, C. M., Castro, P. M. L., & Malcata, F. X. (2012). Metal uptake by microalgae: Underlying mechanisms and practical applications. *Biotechnology Progress, 28*(2), 299–311. https://doi.org/10.1002/btpr.1504.

Mortazavi, S. (2008). Application of membrane separation technology to mitigation of mine effluent and acidic drainage. Natural Resources Canada. MEND. Report No. 3.15.1. https://mend-nedem.org/wp-content/uploads/2013/01/3.15.1.pdf.

Mosley, L. M., Willson, P., Hamilton, B., Butler, G., & Seaman, R. (2015). The capacity of biochar made from common reeds to neutralise pH and remove dissolved metals in acid drainage. *Environmental Science and Pollution Research, 22*(19), 15113–15122. https://doi.org/10.1007/s11356-015-4735-9.

Muedi, K. L., Tendenedzai, J. T., Masindi, V., Haneklaus, N. H., & Brink, H. G. (2025). Facile recovery of polycationic metals from acid mine drainage and their subsequent valorisation for the treatment of municipal wastewater. *Sustainability, 17*(6), 2701. https://doi.org/10.3390/su17062701.

Mulopo, J. (2022). Active physical remediation of acid mine drainage: Technologies review and perspectives. *Journal of Ecological Engineering, 23*(6), 148–163. https://doi.org/10.12911/22998993/147409.

Naidu, T. S., Sheridan, C. M., Doreen, L., & Dyk, V. (2021). Design of acid mine drainage remediation plant report to the water research commission. www.wrc.org.za

Nancucheo, I., Bitencourt, J. A. P., Sahoo, P. K., Alves, J. O., Siqueira, J. O., & Oliveira, G. (2017). Recent developments for remediating acidic mine waters using sulfidogenic bacteria. *BioMed Research International, 2017*. https://doi.org/10.1155/2017/7256582.

Nguegang, N., & Ambushe, A. (2024). Sustainable acid mine drainage treatment: A comprehensive review of passive, combined, and emerging technologies. Environmental Engineering Research https://doi.org/10.4491/eer.2024.592

Nobahar, A., Fitas, E. T., Costa, M. C., & Carlier, J. D. (2024). Acid mine drainage bioremediation using bacteria enriched from the confluence zone between its flow and treated sewage. *International Journal of Environmental Science and Technology*. https://doi.org/10.1007/s13762-024-06160-z.

Novair, S. B., Biglari, Z., Lajayer, B. A., Shu, W., & Price, G. W. (2024). The role of sulphate-reducing bacteria (SRB) in bioremediation of sulphate-rich wastewater: Focus on the source of electron donors. *Process Safety and Environmental Protection, 184*, 190–207. https://doi.org/10.1016/j.psep.2024.01.103.

Oh, S. Y., & Yoon, M. K. (2013). Biochar for treating acid mine drainage. *Environmental Engineering Science, 30*(10), 589–593. https://doi.org/10.1089/ees.2013.0063.

Olawade, D. B., Wada, O. Z., Egbewole, B. I., Fapohunda, O., Ige, A. O., Usman, S. O., & Ajisafe, O. (2024). Metal and metal oxide nanomaterials for heavy metal remediation: Novel approaches for selective, regenerative, and scalable water treatment. *Frontiers in Nanotechnology, 6*. https://doi.org/10.3389/fnano.2024.1466721.

Opoku, P., Gikunoo, E., Arthur, E. K., & Foli, G. (2020). Removal of selected heavy metals and metalloids from an artisanal gold mining site in ghana using indigenous plant species. *Cogent Environmental Science, 6*(1). https://doi.org/10.1080/23311843.2020.1840863.

Orandi, S., Yaghubpur, A., & Sahraei, H. (2007). Influence of AMD on aquatic life at Sar Cheshmeh copper mine. Abstract. *Goldschmidt conference. Cologne, Germany*.

Othman, A., Mamat, C. R., Ibrahim, I., & Rozi, M. S. M. (2023). The mixture of limes in acid mine drainage treatment. *Malaysian Journal of Fundamental and Applied Sciences, 19*, 668–678. doi:10.11113/mjfas.v19n4.2759.

Panda, S., Mishra, S., & Akcil, A. (2016). Bioremediation of acidic mine effluents and the role of sulfidogenic biosystems: A mini-review. *Euro-Mediterranean Journal for Environmental Integration, 1*, 8.

Papirio, S., Villa-Gomez, D. K., Esposito, G., Pirozzi, F., & Lens, P. N. L. (2013). Acid mine drainage treatment in fluidized-bed bioreactors by sulfate-reducing bacteria: A critical review. *Critical Reviews in Environmental Science and Technology, 43*(23), 2545–2580. https://doi.org/10.1080/10643389.2012.694328.

Park, J. H., Ok, Y. S., Kim, S. H., Cho, J. S., Heo, J. S., Delaune, R. D., & Seo, D. C. (2016). Competitive adsorption of heavy metals onto sesame straw biochar in aqueous solutions. *Chemosphere, 142*, 77–83. https://doi.org/10.1016/j.chemosphere.2015.05.093.

Pellera, F. M., Giannis, A., Kalderis, D., Anastasiadou, K., Stegmann, R., Wang, J. Y., & Gidarakos, E. (2012). Adsorption of Cu(II) ions from aqueous solutions on biochars prepared from agricultural by-products. *Journal of Environmental Management, 96*(1), 35–42. https://doi.org/10.1016/j.jenvman.2011.10.010.

Phyo, A. K., Jia, Y., Tan, Q., Sun, H., Liu. , Y., Dong, B., & Ruan, R. (2020). Competitive growth of sulfate reducing bacteria with bioleaching acidophiles for bioremediation of heap bioleaching residue. *International Journal of Environmental Research and Public Health, 17*, 2715. https://doi.org/10.3390/ijerph1708.

Rahardjo, H., Satyanaga, A., Mohamed, H., Ip, C. Y., & Rishi, S. S. (2019). Comparison of soil-water characteristic curves from conventional testing and combination of small-scale centrifuge and dew point methods. *Journal of Geotechnical and Geological Engineering, 37*(2), 659–672.

Rambabu, K., Banat, F., Pham, Q. M., Ho, S. H., Ren, N. Q., & Show, P. L. (2020). Biological remediation of acid mine drainage: Review of past trends and current outlook. *Environmental Science and Ecotechnology, 2*. https://doi.org/10.1016/j.ese.2020.100024.

Ramla, B., & Sheridan, C. (2015). The potential utilisation of indigenous South African grasses for acid mine drainage remediation. *Water SA, 41*(2), 247–252. https://doi.org/10.4314/wsa.v41i2.10.

Roy, A., & Bhattacharya, J. (2015). *Nanotechnology in industrial wastewater treatment*. IWA Publishing.

Rose, A.W., 2010. Advances in passive treatment of coal mine drainage 1998-2009. 2010 National Meeting of the American Society of Mining and Reclamation, Pittsburgh, PA Bridging Reclamation, Science and the Community June 5 - 11, 2010. R.I. Barnhisel (Ed.) Published by ASMR, 3134 Montavesta Rd., Lexington, KY 40502. https://doi.org/10.21000/jasmr10010847.

RoyChowdhury, A., Sarkar, D., & Datta, R. (2015). Remediation of acid mine drainage-impacted water. *Current Pollution Reports, 1*(3), 131–141. Issue https://doi.org/10.1007/s40726-015-0011-3 .

Ryu, S., Naidu, G., Hasan Johir, M. A., Choi, Y., Jeong, S., & Vigneswaran, S. (2019). Acid mine drainage treatment by integrated submerged membrane distillation–sorption system. *Chemosphere, 218*, 955–965. https://doi.org/10.1016/j.chemosphere.2018.11.153.

Sanchez-Andrea, I., Stams, A. J. M., & Hedrich, S. (2015). *Desulfosporosinus acididurans* sp. nov.: An acidophilic sulfate-reducing bacterium isolated from acidic sediments. *Extremophiles, 19*, 39–47. https://doi.org/10.1007/s00792-014-0701-6.

Schlüter, S., Lucas, M., Grosz, B., Ippisch, O., Zawallich, J., He, H., Dechow, R., Kraus, D., Blagodatsky, S., Senbayram, M., Kravchenko, A., Vogel, H.-J., & Well, R. (2025). The anaerobic soil volume as a controlling factor of denitrification: A review. *Biology and Fertility of Soils, 61*, 343–365. https://doi.org/10.1007/s00374-024-01819-8.

Sharma, K., Shah, G., Singh, H., Bhatt, U., Singhal, K., & Soni, V. (2024). Advancements in natural remediation management techniques for oil spills: Challenges, innovations, and future directions. *Environmental Pollution and Management, 1*, 128–146. https://doi.org/10.1016/j.epm.2024.08.003.

Shi, J., Fan, X., Tsang, D. C. W., Wang, F., Shen, Z., Hou, D., & Alessi, D. S. (2019). Removal of lead by rice husk biochars produced at different temperatures and implications for their environmental utilizations. *Chemosphere, 235*, 825–831. https://doi.org/10.1016/j.chemosphere.2019.06.237.

Shirvanimoghaddam, K., Czech, B., Abdikheibari, S., Brodie, G., Konczak, ´M., Krzyszczak, A., Al-Othman, A., & Naebe, M. (2022). Microwave synthesis of biochar for environmental applications. *Journal of Analytical and Applied Pyrolysis, 161*, 105415. https://doi.org/10.1016/j.jaap.2021.105415.

Tan, X., Liu, Y., Zeng, G., Wang, X., Hu, X., Gu, Y., & Yang, Z. (2015). Application of biochar for the removal of pollutants from aqueous solutions. *Chemosphere, 125*, 70–85. https://doi.org/10.1016/j.chemosphere.2014.12.058.

Tong, X., Li, J., Yuan, J., & Xu, R. (2011). Adsorption of Cu(II) by biochars generated from three crop straws. *Chemical Engineering Journal, 172*, 828–834. https://doi.org/10.1016/j.cej.2011.06.069.

Trakal, L., Bingol, ¨D., Pohoˇrelý, M., Hruˇska, M., & Kom´ arek, M. (2014). Geochemical and spectroscopic investigations of Cd and Pb sorption mechanisms on contrasting biochars: Engineering implications. *Bioresource Technology, 171*, 442–451. https://doi.org/10.1016/j.biortech.2014.08.108.

Tran, H. N., You, S. J., & Chao, H. P. (2016). Effect of pyrolysis temperatures and times on the adsorption of cadmium onto orange peel derived biochar. *Waste Management and Research, 34*(2), 129–138. https://doi.org/10.1177/0734242X15615698.

Tran, T. K., Chiu, K. F., Lin, C. Y., & Leu, H. J. (2017). Electrochemical treatment of wastewater: Selectivity of the heavy metals removal process. *International Journal of Hydrogen Energy, 42*, 27741–27748.

Trumm, D. (2010). Selection of active and passive treatment systems for AMD—Flow charts for New Zealand conditions. *New Zealand Journal of Geology and Geophysics, 53*(2–3), 195–210. https://doi.org/10.1080/00288306.2010.500715.

Urrutia, C., Yañez-Mansilla, E., & Jeison, D. (2019). Bioremoval of heavy metals from metal mine tailings water using microalgae biomass. *Algal Research*, 43. https://doi.org/10.1016/j.algal.2019.101659.

van der Bruggen, B., Vandecasteele, C., van Gestel, T., Doyen, W., & Leysen, R. (2003). A review of pressure-driven membrane processes in wastewater treatment and drinking water production. *Environmental Progress, 22*(1), 46–56.

Van Epps, A. (2006). *Phytoremediation of petroleum hydrocarbons*. Environmental Protection Agency.

Van Hille, R., Boshoff, G. A., Rose, P. D., & Duncan, J. R. (1999). A continuous process for the biological treatment of heavy metal contaminated acid mine water. *Resources, Conservation & Recycling., 27*, 157–167.

Wang, H., Gao, B., Wang, S., Fang, J., Xue, Y., & Yang, K. (2015). Removal of Pb(II), Cu(II), and Cd(II) from aqueous solutions by biochar derived from KMnO4 treated hickory wood. *Bioresource Technology, 197*, 356–362. https://doi.org/10.1016/j.biortech.2015.08.132.

Wang, J., Zhang, S., He, C., She, Z., Pan, X., Li, Y., Shao, R., Shi, Q., & Yue, Z. (2020a). Source identification and component characterization of dissolved organic matter in an acid mine drainage reservoir. *Science of the Total Environment, 739*. https://doi.org/10.1016/j.scitotenv.2020.139732.

Wang, X., Ning, Z., Sun, W., Liu, H., & Li, B. (2020b). Energy and environmental impact assessment of a passive remediation bioreactor for antimony-rich mine drainage. *Environmental Science and Pollution Research, 27*(28), 3504035050. https://doi.org/10.1007/s11356-020-09816-8.

Wang, Z., Xu, Y., Zhang, Z., & Zhang, Y. (2021b). Review: Acid mine drainage (AMD) in abandoned coal mines of Shanxi, China. *Water (Switzerland), 13*(1). https://doi.org/10.3390/w13010008.

Wei, D., Li, B., Luo, L., Zheng, Y., Huang, L., Zhang, J., Yang, Y., & Huang, H. (2020). Simultaneous adsorption and oxidation of antimonite onto nano zero-valent iron sludge-based biochar: Indispensable role of reactive oxygen species and redox-active moieties. *Journal of Hazardous Materials., 391*, 122057. https://doi.org/10.1016/j.jhazmat.2020.122057.

Wu, J., Huang, D., Liu, X., Meng, J., Tang, C., & Xu, J. (2018). Remediation of As(III) and Cd (II) co-contamination and its mechanism in aqueous systems by a novel calciumbased magnetic biochar. *Journal of Hazardous Materials, 348*, 10–19. https://doi.org/10.1016/j.jhazmat.2018.01.011.

Wu, J., Wang, T., Wang, J., Zhang, Y., & Pan, W.-P. (2021). A novel modified method for the efficient removal of Pb and Cd from wastewater by biochar: Enhanced the ion exchange and precipitation capacity. *Science of the Total Environment, 754*, 142150. https://doi.org/10.1016/j.scitotenv.2020.142150.

Xiong, Q., Zhang, L., Xun, A., & Du, J. (2025). Review on the application research of covers with capillary barrier effects. *Journal of Research in Science and Engineering,* 7(1), 125–139. https://doi.org/10.53469/jrse.2025.07(01).19.

Xu, X., Cao, X., Zhao, L., Wang, H., Yu, H., & Gao, B. (2013). Removal of Cu, Zn and Cd from aqueous solutions by the dairy manure-derived biochar. *Environmental Science and Pollution Research, 20*, 358–368. https://doi.org/10.1007/s11356-012-0873-5.

Yang, T., Xu, Y., Huang, Q., Sun, Y., Liang, X., Wang, L., Qin, X., & Zhao, L. (2021). Adsorption characteristics and the removal mechanism of two novel Fe-Zn composite modified biochar for Cd(II) in water. *Bioresource Technology*, 333. https://doi.org/10.1016/j.biortech.2021.125078.

Yang, Y., Zhang, Y., Wang, G., Yang, Z., Xian, J., Yang, Y., Li, T., Pu, Y., Jia, Y., Li, Y., Cheng, Z., Zhang, S., & Xu, X. (2021b). Adsorption and reduction of Cr(VI) by a novel nanoscale FeS/chitosan/biochar composite from aqueous solution. *Journal of Environmental Chemical Engineering, 9*, 105407. https://doi.org/10.1016/j.jece.2021.105407.

Yoon, K., Cho, D. W., Tsang, D. C. W., Bolan, N., Rinklebe, J., & Song, H. (2017). Fabrication of engineered biochar from paper mill sludge and its application into removal of arsenic and cadmium in acidic water. *Bioresource Technology, 246*, 69–75. https://doi.org/10.1016/j.biortech.2017.07.020.

Yu, J. W., Kim, J. S., Li, X., Jong, Y. C., Kim, K. H., & Ryang, G. I. (2022). Water quality forecasting based on data decomposition, fuzzy clustering and deep learning neural network. *Environmental Pollution, 303*. https://doi.org/10.1016/j.envpol.2022.119136.

Zagury, G.J., & Neculita, C. (2007). Passive treatment of acid mine drainage in bioreactors: Short review, applications, and research needs. Ottawa Geo. Conference 2007, 1439-1445p.

Zhao, Y. (2018). *Chapter 3 - Biological Treatment Processes for Leachate: Pollution Control Technology for Leachate from Municipal Solid Waste*: 185-324. Butterworth-Heinemann https://doi.org/10.1016/B978-0-12-815813-5.00003-6.

CHAPTER FIVE

Real-time AI monitoring of acid mine drainage

5.1 Artificial intelligence–driven system for monitoring acid mine drainage

Unlike in Chapters 3 and 4 where treatment methods for AMD have been discussed, in this chapter, the main focus is on monitoring of AMD. The discussion is pegged on artificial intelligence (AI) rooted in the challenges and complexities encountered when detecting, predicting and monitoring AMD using traditional methods—the complexities are a function of both environmental and geological factors (Mardonova et al., 2025). This chapter therefore presents alternatives that increase efficiency of AMD monitoring without compromising quality outputs. One of the alternatives lies in adopting and using advanced technology such as machine learning (ML) and AI (More & Wolkersdorfer, 2023).

Traditional AMD monitoring relies heavily on periodic manual sampling, laboratory chemical analysis and static threshold-based alert systems. Such approaches are time-consuming, expensive and incapable of capturing the rapid fluctuations in water chemistry that characterize active mining environments. In contrast, AI and ML offer dynamic, data-driven alternatives that can process large volumes of sensor data in near real-time, identify complex nonlinear relationships between environmental variables and generate reliable predictive forecasts of water quality deterioration before critical thresholds are breached (Hasrod et al., 2024; Mardonova et al., 2025; More & Wolkersdorfer, 2023).

5.1.1 Fundamentals of AI and ML in AMD

Artificial intelligence and ML are achieved through training of datasets using algorithms (step-by-step sets of instructions or rules designed to solve a specific problem or perform a task). In the context of AMD, training using these algorithms leads to the generation of predictive models (Hasrod et al., 2024; Mardonova et al., 2025; More & Wolkersdorfer, 2023) capable of detecting

Acid Mine Drainage: Sustainable Mining Practices for Mitigating Environmental Impact
DOI: https://doi.org/10.1016/B978-0-443-45107-2.00006-X

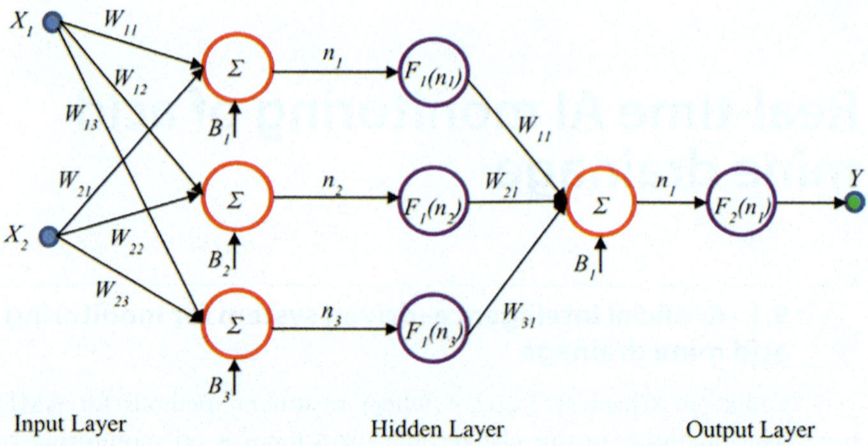

Figure 5.1 Prediction of sulphate concentration using LSTM algorithms. Future sulfate concentration in AMD generated by LSTM algorithms (Sakala, Novhe, & Vadapalli, 2019). *AMD*, Acid mine drainage; *LSTM*, long short-term memory. *(Reproduced with permission from Ozel et al. (2020)).*

and monitoring AMD dynamics (Mardonova et al., 2025; Popescu et al., 2024). Parameters required for dataset training include pH, metal concentration, hydrology, flow rate, and climatic conditions such as temperature and rainfall.

Training aimed at obtaining a predictive model is achieved through algorithms such as artificial neural networks (ANNs) and long short-term memory (LSTM) networks (Mardonova et al., 2025; Sakala, Novhe, & Vadapalli, 2019). The training of ANN or LSTM algorithms involves the use of layered architectures (Fig. 5.1) comprising input layers, hidden layers, and output layers (Ozel et al., 2020).

Input layers receive raw sensor observations; hidden layers apply nonlinear transformations through weighted connections and activation functions; and output layers generate the predicted values of interest, such as metal concentrations or pH forecasts. The depth and breadth of hidden layers determine model complexity and hyperparameter optimization including learning rate, dropout rate, and number of training epochs that governs overall training performance.

The datasets are collected with the assistance of automated sensors that measure pH levels, dissolved oxygen (DO), electrical conductivity (EC), temperature, turbidity, flow rate and metal ion concentrations (Akcil & Koldas, 2006). These sensors are typically deployed at strategic monitoring points along drainage channels, in tailings ponds and at seepage

Table 5.1 Algorithms used to train datasets in various AI models.

Algorithm type	Key features	Data requirements	Prediction model purpose
ANN	CL, RG, CLS Time series prediction	Data-intensive, poor performance on small data sets	Heavy metals water chemistry and seepage flow
LSTM	Time series prediction	Sequential data, perform best with clean and normalized data	Seepage flow and chemistry of heavy metals
ANFIS	Integrated FL and NN	Huge data volume, scaling and noise free	Heavy metals
GRNN	NLR, CL, RG	Relatively large data storage memory required	Heavy metals
SVM	CL, RG, NLR	Slow in large datasets. Sensitive outliers	Heavy metals
BPNN	CL, prediction and optimization	Data intensive, potential overfitting	Heavy metals
M5 prime regression model	Handling continuous class learning problems	Labeled numeric data with pre-processing	Heavy metals
KNN	Instance-based learning.	Labeled numeric data with pre-processing and distance in metric compatibility	Heavy metals

AI, artificial intelligence; *ANFIS*, adaptive neuro-fuzzy inference system; *ANN*, artificial neural network; *BPNN*, back propagation neural network; *GRNN*, general regression neural network; *KNN*, K-nearest neighbor; *LSTM*, long short-term memory; *SVM*, support vector machine; CL, Continue learning; CLS, Classification system; FL, Federated leaning; NLR, Nonlinear regression; NN, Neural network; RG, Regression.
From Mardonova et al. (2025).

outfalls, providing continuous high-frequency measurements at intervals ranging from seconds to minutes depending on site requirements. The feedback on proposed decisions is then provided for immediate remedial action (Mardonova et al., 2025). Other common algorithms beyond ANN and LSTM include random forest (RF), adaptive neuro-fuzzy inference system (ANFIS), K-nearest neighbor, general regression neural network, back propagation neural network, and support vector machine (SVM) (Table 5.1). Algorithms may be trained to analyze large datasets collected from AMD sites. The accuracy of a model generated will depend on the availability of sufficiently large datasets—a prerequisite for obtaining reliable patterns—as well as noise correction, data labeling and feature engineering procedures (Mardonova et al., 2025). Commonly, ANN, SVM, and ANFIS have been trained to generate predictive models for different heavy metals, such as Cu, Zn, Cd, Fe, Mn, Pb, Ni, and As in AMD (Mardonova et al., 2025; Rooki et al., 2011).

The accuracy and efficiency of a model is validated by checking multiple indicator metrics, such as correlation coefficient (*R*), coefficient determination (R^2), root mean square error (RMSE), mean absolute error (MAE), mean square error, standard error, variance accounted for and other statistical metrics indicators (Table 5.2).

To validate a given model, metrics are used to test the successfulness in predictability of different heavy metals in AMD (e.g., Agah & Soleiman-pourmoghadam, 2020; Aryafar et al., 2012; Betrie et al., 2014; Rooki et al., 2011). Betrie et al. (2014) use ANN and SVM to generate prediction model of AMD chemistry of heavy metals using data collected for 25 years in Cu–Mo–Au–Ag–Re mine in British Columbia. Data collected were pH, conductivity (μS/cm), acidity (mg/L), alkalinity ($CaCO_3$ mg/L), sulfate (mg/L), flow and concentration of Cu, Cd, Zn, Ca, Mg and Al (mg/L). During algorithm training, missing and outlier values were considered to avoid overlapping of information. Maximum criterion information (a nonlinear correlation function) was used. The validation of generated model by ANN and SVM was done by statistical techniques (i.e., RMSE, MAE, RRSE and RE) where for both techniques small values indicate better performance (Table 5.2). They concluded that ANN prediction model was better enveloped for Cu and Zn compared to the model generated by SVM for both Cu and Zn, while for Cd, both ANN and SVM models were not good, because more than 90% of the Cd concentration data were below analytical detection limit of 0.05 mg/L (Fig. 5.2).

Artificial intelligence has demonstrated outstanding capability to generate predictive models in AMD. For instance, Sakala et al. (2019) apply LSTM to generate a predictive model that show future concentration of sulfate in AMD. Using LSTM algorithm, they trained historical datasets such as temperature, water level, rainfall, pH and sulfate and were able to establish future corresponding values of temperature, water level, rainfall, pH and sulfate. Despite some uncertainty behind AI model raised due to errors, limited sample size for data (since it requires huge data set to recognize and develop pattern), yet, the use of AI is outstanding and remains a promising tool for AMD prediction and monitoring. More and Wolkersdorfer (2023) conducted research in South Africa by employing

Table 5.2 Metric indicators used in validating accuracy and efficiency of a predictive model.

Metric indicator	Description	Calculation	Range	Interpretation	References
MAE	Average error between predicted and calculated	Mean absolute differences	[0,N]	Lower value indicates better performance	Yang (2023)
MSE	Mean squared difference between prediction and actual values	Mean squared difference	[0,N]	Lower value indicates better performance	Wu (2022)
RMSE	Average magnitude of error between prediction and actual values	Square root of mean square difference	[0,N]	Lower value indicates better performance	Yang (2023), Wu (2022)
R^2	Proportion of variance in dependent variable predictable from independent variable	Calculated from regression output	[0,1]	Higher R^2 suggest better fit	Bhagat et al. (2021)

(continued on next page)

Table 5.2 Metric indicators used in validating accuracy and efficiency of a predictive model—cont'd

Metric indicator	Description	Calculation	Range	Interpretation	References
R	Strength and Direction of Linear relationship between two variables	Calculated from co-variance and standard deviation	[-1,1]	Closer to 1 or 1 means stronger correlation	Alizamir et al. (2019)
MAPE	Percentage difference between predicted and actual values	Mean of absolute percentage difference	[0, N]	Lower values indicate better accuracy	Lu and Ma (2020), Yu et al. (2022)
SE	Accuracy of sample estimates and their variability from population parameter	Standard deviation divided by square root of sample size	[0,N]	Lower values better performance	Magesh et al. (2016)
VAF	Their variance in a models prediction	Explained variance as percentage of total variance in data	[0%, 100%]	Higher values indicate better performance	Agah and Soleiman-pour-moghadam (2020)

MAE, mean absolute error; *MSE*, mean square error; *RMSE*, root mean square error; *SE*, standard error; *VAF*, variance accounted forReproduced with permission from Mardonova et al. (2025).

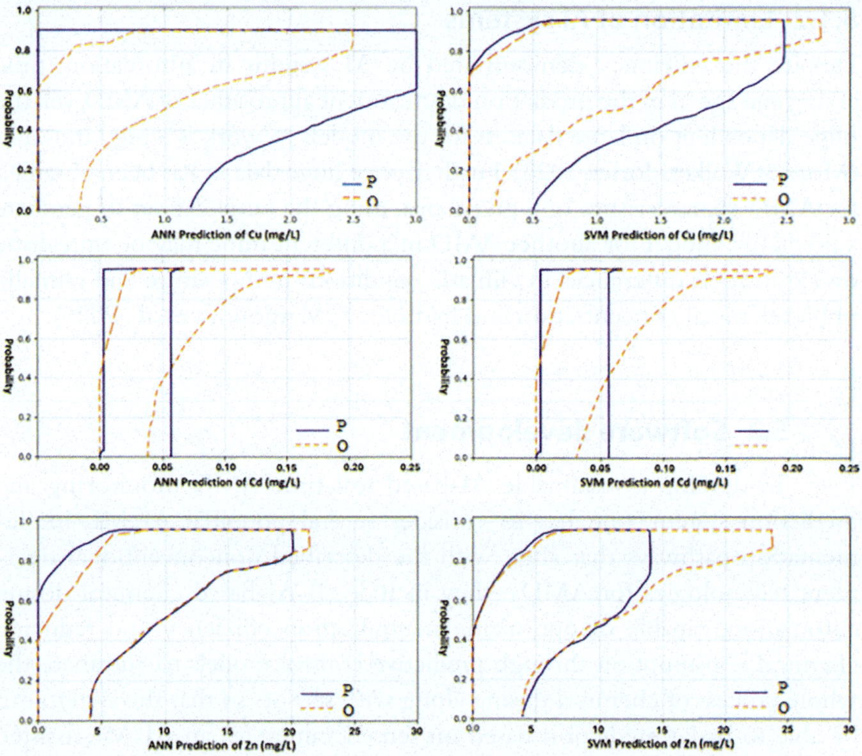

Figure 5.2 Predicted and observed models for Cu, Cd, and Zn in (mg/L) by ANN and SVM algorithms. *ANN*, Artificial neural network; *O*, observed; *P*, predicted; *SVM*, support vector machine. *(Reproduced with permission from Betrie et al. (2014)).*

several ML algorithms to predict a model for AMD. Algorithm of deep neural network, ANN, RF and gradient boosting regression were used through the analyses of parameters such pH, concentration of Fe, Al, and SO_4^{2-}. Similar studies were also conducted by More and Wolkersdorfer (2022) who combined ANN and SVM algorithms to generate predictive model of AMD by feeding parameters such as mine depth, water level, rainfall, dissolved solids (TDS), EC and pH. Using meteorological data, Zhang (2023) also researched on AI on the prediction capability of AMD. The model of Zhang (2023) relied on decision tree for classification and LSTM algorithm for prediction drainage flow rate. The outcome was 91% for classification which defined improved flow rate performance in the model.

5.1.2 Limitation of AI systems

Despite the efficiency demonstrated by AI systems in automation, time saving and speed in the prediction on treatment approaches of AMD, yet, the huge dependent on large datasets for the models to work is a big limitation (More & Wolkersdorfer, 2023). Further, once huge dataset has been obtained for AMD characteristics of a given area, using the same dataset to generate a predictive model for another AMD in a different mine may be unrealistic owing to their differences in climatic conditions (temperature and rainfall), pH level, metal concentration and hydrology (Mardonova et al., 2025).

5.2 Software development

To develop a sustainable AI-based real-time AMD monitoring and prediction system, one has to consider several potential AI/ML implementation pathways that align with the described nonconventional treatment technologies for AMD—they include (1) AI-based chemical dosing automation, capable of optimizing neutralization efficiency and reducing chemical consumption through predictive control models to automate the whole process of chemical dosing along with sensors so that the system can be able to make prediction based on sensed values; (2) an AI/ML-assisted decision support system to compare treatment methods such as microbial remediation, nanotechnology, membrane filtration, biochar adsorption, and phytoremediation, depending on specific AMD composition and site conditions including budget; (3) a digital twin simulation environment for modeling AMD evolution, treatment performance and system behavior under different operational scenarios; (4) computer-vision solutions using drone imagery to detect emerging AMD sources, tailings seepage zones, and landscape degradation; and (5) AI models for long-term prediction of metal concentration trends, using machine learning methods such as ANN, LSTM, SVM, and RF as highlighted in Table 5.1. In this book, the authors chose to implement an AI-based real-time AMD monitoring and prediction system. This approach is technically feasible because it leverages accessible data streams (pH, temperature, flow rate, sulfate and metal concentrations) (Hasrod et al., 2024; Mardonova et al., 2025). Other data, required other than pH, temperature, flow rate, sulfate, and metal concentration, include EC, DO, turbidity, oxidation-reduction potential, TDS, rainfall intensity, air temperature, barometric pressure, and where available online metal concentrations for Fe, Mn, Cu, Zn, and Al. These are integrated with predictive ML models to forecast acidity levels, metal spikes and treatment needs. The developed system operates on a modular architecture that integrates IoT sensors, time-series models, anomaly detection algorithms and a decision-support dashboard. This makes it practical to deploy, to scale it across sites and capable of delivering immediate operational values without requiring specialized treatment infrastructure or complex experimental setups. On the live system created, it allows simulation of sensor (IoT sensor) data in real time to show performance—the system shall be connected by sensors for good results. This approach, adopted in this book, bridges scientific research with modern digital innovation by adding an AI-driven monitoring layer—the book extends beyond describing treatment methods to demonstrating how emerging technologies can transform the management of AMD from being merely descriptive to being an actionable tool for the industry, academia, and environmental agencies. To make this software package more relevant and applicable, detailed technology stacks have been highlighted. Scholars can follow them to achieve similar or better results.

This system is a multi-model system, that is, it applies not just a single ML model but several of them, such as Webapp—React, Backend—Microservice architecture which will combine Python and Node, MQTT broker (Mosquitto) for lightweight telemetry, PostgreSQL with TimescaleDB extension, JWT-based auth for APIs, role-based access for dashboard (admin/operator/viewer) and ML Algorithm—as indicated hereunder.

1. LSTM (TensorFlow/Keras)—Short-term real-time forecasting
   Purpose: Forecasts pH, sulfate, and metal levels 1–72 hours ahead. Dataset needed (time-series dataset)—a continuous, timestamped multivariate time-series dataset. Data from at least three to six months are needed to for best performance and accuracy. The format used is CSV or Parquet. Columns used are timestamp, pH, temp, conductivity, flow, turbidity, rainfall, air_temp and so on.
2. XGBoost Regression—Medium-term metal concentration prediction.
   Purpose: Predicts Fe, Cu, Zn, Mn, Al, etc., for 1–30 days ahead. Dataset needed (lab + engineered features)—XGBoost needs tabular data—this would require presence of (a) lab measured heavy metal datasets and (b) matching sensor data. This is needed to give possibility of modeling relationship between real-time conditions and metals—it aims at aligning sensor data with lab timestamps. The laboratory rows are merged with (i) average sensor data from the previous 1, 6, 12, and 24 hour data, (ii) rainfall of last 24 and 72 hours, (iii) flow rate trends, (iv) pH variability, and (v) conductivity patterns; (c) engineered features dataset. In the latter, tsfresh

will help together with python automation scripts have helped to capture (i) rolling_mean_pH_24h, (ii) rolling_std_pH_24h, (iii) delta_pH_6h, (iv) rolling_mean_conductivity_24h, (v) rolling_sum_rainfall_24h, (vi) rolling_sum_rainfall_72h, (vii) flow_rate_gradient, and (viii) turbidity_changes. Minimum dataset is 50–200 lab samples matched with sensor patterns.

3. IsolationForest + Ruptures—Anomaly detection dataset.
Purpose: Detects sudden unusual behavior in pH, flow, conductivity, and metals. Dataset used here are (a) normal (baseline) time-series sensor data for pH, flow, conductivity, temp, and turbidity, and other similar data can be accepted in (b) rainfall and (c) derived statistics (rolling means). The data should cover both normal conditions (majority) and known anomaly periods (optional but helpful) for pH drops, heavy rainfall causing spikes, sensor failure events and discharge events, and (d) labels (optional). Labels are not required but useful. Label = normal/anomaly. For good results, data spanning 1–3 months of continuous time-series are needed.
4. LightGBM quantile regression prediction intervals.
Its main purpose is to provide confidence bounds around predictions to show prediction accuracy. It is bound to be around specific limits. Dataset needed are same as for number 2 (i.e., same dataset used for XGBoost and LSTM) but preprocessed into a tabular form with the following variables: (i) rolling_window_features, (ii) lag features (1, 6, 12, and 24 hours), (iii) rainfall aggregates, (iv) flow aggregates, and (v) pH volatility. Training labels are (i) future pH, (ii) future sulfate, (iii) future metals, and (iv) quantiles: 0.1, 0.5 and 0.9 (ground truth values, no extra data needed). How it works is that LightGBM uses the same data as core models but trains on different loss functions.
5. the tsfresh
This is used for feature Extraction Dataset. Purpose: Automatically create hundreds of engineered features from the data. These are needed in order that models established can understand the data input. Database needed are a clean time series with columns id (site ID), time (timestamp), and sensor values including pH, temperature, conductivity, flow rate, turbidity, and rainfall. The computed features are subsequently filtered by relevance using built-in hypothesis testing to retain only those significantly correlated with the prediction target, reducing dimensionality, and improving model generalization.

Table 5.3 Technology stacks needed for AI software development.

Layer	Technology	Role and justification
Frontend	React (TypeScript)	Responsive single-page application delivering the real-time dashboard; supports role-based views (admin, operator, viewer) and live chart rendering via WebSocket subscriptions
Backend	Python (FastAPI) + Node.js (Express)	Microservice architecture: Python handles ML inference endpoints and data preprocessing pipelines; Node.js manages REST APIs, JWT authentication, user management, and WebSocket broadcast to the frontend
IoT/telemetry	MQTT (Mosquitto broker)	Lightweight publish-subscribe protocol optimized for constrained field sensors; decouples sensor hardware from backend processing and enables multi-site data ingestion with minimal bandwidth
Database	PostgreSQL + TimescaleDB	TimescaleDB extends PostgreSQL with automatic partitioning of time-series sensor data, compression, and continuous aggregate views; enables sub-second queries across months of high-frequency readings
ML training and inference	TensorFlow/Keras, XGBoost, LightGBM, scikit-learn, tsfresh	Each model handles a distinct prediction horizon and task: LSTM for 1–72 hours forecasting; XGBoost for 1–30 day metal prediction; LightGBM for quantile confidence intervals; IsolationForest + Ruptures for anomaly detection; tsfresh for automated feature engineering
Authentication and access	JWT + RBAC	JSON Web Token authentication secures all API endpoints; role-based access control restricts dashboard capabilities by user type, ensuring data integrity and regulatory audit readiness
Containerization and deployment	Docker + Docker Compose	Each service runs in an isolated container; Docker Compose orchestrates local and cloud deployments, enabling reproducible scale-out across multiple mine sites

AI, artificial intelligence; *ML*, machine learning.

5.2.1 Technology stack

To develop this kind of software, one would need more relevant and applicable detailed technology stacks—the most important are frontend, backend, IoT/telemetry, database, ML training and inference, authentication and access, containerization and deployment (Table 5.3). Scholars and practitioners can follow them to achieve similar or improved results.

5.2.2 Real-word impact

This system has the potential to significantly benefit the mining and environmental management sector. Real-time prediction reduces treatment costs, prevents ecological damage before it occurs, improves compliance with regulatory standards and supports the sustainable operation of mines. Therefore, enabling continuous insight into water chemistry and providing intelligent recommendations, the solution can reshape how AMD is handled globally. Its scalability and data-driven nature make it a profitable innovation for solution providers, monitoring firms and environmental consultancies, while its preventive capabilities position it as a world-changing tool capable of reducing one of the most persistent environmental challenges in mining.

Future development trajectories include the integration of satellite multispectral remote sensing data for large-scale tailings monitoring, the adoption of federated learning approaches that allow models to be trained across multiple mine sites without centralizing sensitive operational data and the incorporation of regulatory forecasting modules capable of comparing predicted future concentrations against jurisdiction-specific discharge standards and automatically generating compliance reports. These advances will further increase the system's operational relevance for mine operators navigating evolving environmental legislation (Mardonova et al., 2025; Yang, 2023). The link to the software developed www.amdtoolkit.com.

References

Agah, A., & Soleimanpourmoghadam, N. (2020). Design and implementation of heavy metal prediction in acid mine drainage using multi-output adaptive neuro-fuzzy inference systems (ANFIS) – A case study. *International Journal of Mining and Geological Engineering, 54*(1), 59–64. https://doi.org/10.22059/IJMGE.2019.278558.594794.

Akcil, A., & Koldas, S. (2006). Acid mine drainage (AMD): Causes, treatment and case studies. *Journal of Cleaner Production, 14*, 1139–1145.

Alizamir, M., Sobhanardakani, S., & Hasanalipour Shahrabadi, A. (2019). Prediction of heavy metals concentration in the groundwater resources in Razan Plain: Extreme learning machine vs. artificial neural network and multivariate adaptive regression spline. *Annals of Military and Health Sciences Research, 17*(4). https://doi.org/10.5812/amh.98554.

Aryafar, A., Gholami, R., Rooki, R., & Doulati Ardejani, F. (2012). Heavy metal pollution assessment using support vector machine in the Shur River, Sarcheshmeh copper mine. *Iran. Environmental Earth Sciences, 67*(4), 1191–1199. https://doi.org/10.1007/s12665-012-1565-7.

Betrie, G. D., Sadiq, R., Morin, K. A., & Tesfamariam, S. (2014). Uncertainty quantification and integration of machine learning techniques for predicting acid rock drainage chemistry: A probability bounds approach. *Science of the Total Environment, 490*, 182–190. https://doi.org/10.1016/j.scitotenv.2014.04.125.

Bhagat, S. K., Tung, T. M., & Yaseen, Z. M. (2021). Heavy metal contamination prediction using ensemble model: Case study of Bay sedimentation, Australia. *Journal of Hazardous Materials, 403*. https://doi.org/10.1016/j.jhazmat.2020.123492.

Hasrod, T., Nuapia, Y. B., & Tutu, H. (2024). Comparison of individual and ensemble machine learning models for prediction of sulphate levels in untreated and treated acid mine drainage. *Environmental Monitoring and Assessment, 196*(4), 332. article https://doi.org/10.1007/s10661-024-12467-8 .

Lu, H., & Ma, X. (2020). Hybrid decision tree-based machine learning models for short-term water quality prediction. *Chemosphere, 249*. https://doi.org/10.1016/j.chemosphere.2020.126169.

Magesh, N. S., Chandrasekar, N., & Elango, L. (2016). Occurrence and distribution of fluoride in the groundwater of the Tamiraparani River basin, South India: A geostatistical modeling approach. *Environmental Earth Sciences, 75*(23). https://doi.org/10.1007/s12665-016-6293-y.

Mardonova, M., Shahid, M. K., Abbassi, R., Lim, J. W., Paudel, S. R., & Mainali, B. (2025). Chapter 2. Advances in acid mine drainage management through artificial intelligence. In *Artificial intelligence in future mining* (pp. 77–177). Elsevier. https://doi.org/10.1016/B978-0-443-28911-8.00002-0.

More, K. S., & Wolkersdorfer, C. (2022). Predicting and forecasting mine water parameters using a hybrid intelligent system. *Water Resources Management, 36*(8), 2813–2826. https://doi.org/10.1007/s11269-022-03177-2.

More, K. S., & Wolkersdorfer, C. (2023). Application of machine learning algorithms for nonlinear system forecasting through analytics—A case study with mining influenced water data. *Water Resources and Industry, 29*. https://doi.org/10.1016/j.wri.2023.100209.

Ozel, H. U., Gemici, B. T., Gemici, E., Ozel, H. B., Cetin, M., & Sevik, H. (2020). Application of artificial neural networks to predict the heavy metal contamination in the Bartin River. *Environmental Science and Pollution Research, 27*(34), 42495–42512. https://doi.org/10.1007/s11356-020-10156-w.

Popescu, S. M., Mansoor, S., Wani, O. A., Kumar, S. S., Sharma, V., Sharma, A., Arya, V. M., Kirkham, M. B., Hou, D., Bolan, N., & Chung, Y. S. (2024). Artificial intelligence and IoT driven technologies for environmental pollution monitoring and management. *Frontiers in Environmental Science, 12*. https://doi.org/10.3389/fenvs.2024.1336088.

Rooki, R., Ardejani, F. D., Aryafar, A., & Asadi, A. B. (2011). Prediction of heavy metals in acid mine drainage using artificial neural network from the Shur River of the Sarcheshmeh porphyry copper mine, Southeast Iran. *Environmental Earth Sciences, 64*(5), 1303–1316. https://doi.org/10.1007/s12665-011-0948-5.

Sakala, E., Novhe, O., & Vadapalli, V. R. K. (2019). Application of artificial intelligence (AI) to predict mine water quality: A case study in South Africa. In Ch. Wolkersdorfer, E. Khayrulina, S. Polyakova, & A. Bogush (Eds.), *IMWA 2019 "Mine Water: Technological and Ecological Challenges* (pp. 140–145). International Mine Water Association.

Wu, Z., Duan, H., Li, K., & Ye, L. (2022). A comprehensive carbon footprint analysis of different wastewater treatment plant configurations. *Environmental Research, 214*. https://doi.org/10.1016/j.envres.2022.113818.

Yang, S., Lian, H., Xu, B., Thanh, H. V., Chen, W., Yin, H., & Dai, Z. (2023). Application of robust deep learning models to predict mine water inflow: Implication for groundwater environment management. *Science of the Total Environment, 871.* https://doi.org/10.1016/j.scitotenv.2023.162056.

Yu, J. W., Kim, J. S., Li, X., Jong, Y. C., Kim, K. H., & Ryang, G. I. (2022). Water quality forecasting based on data decomposition, fuzzy clustering and deep learning neural network. Environmental Pollution, 303. Available from https://doi.org/10.1016/j.envpol.2022.119136, https://www.journals.elsevier.com/environmental-pollution.

Zhang, Ma, L., & Liu. (2023). A machine learning approach for prediction of the quantity of mine waste rock drainage in areas with spring freshet. Minerals., 13. Available from https://doi.org/10.3390/min13030376.

CHAPTER SIX

Policy, governance and community engagement

Mineral policies are commonly the source of laws governing exploration and mining activities in any country (Deberdt & Letourneau, 2025; Deberdt et al., 2025). Policies are broad statements that establish a framework for decision-making, ensuring that an organization's or government's actions remain consistent with its long-term vision and goals. Depending on the sectoral vision and mission, these policies often integrate elements such as value addition and environmental protection (Cheang, 2026; Christopher Ulua & Uju, 2021; Jonker-Hoffrén & Sippola, 2024). In the context of acid mine drainage (AMD), mineral and environmental policies are critical because they provide the mandatory framework for environmental protection, long-term liability and technical prevention. Without good policies, AMD management often defaults to reactive, short-term fixes rather than sustainable, proactive solutions (Masindi et al., 2022; Naidoo, 2017).

Policies and legislation governing AMD in countries with a long history of mining are discussed in Jacobs and Testa (2014), Naidoo (2017) and Simate (2021). While Jacobs and Testa (2014) and Simate (2021) provide a foundational analysis of AMD policies in countries with legacy mining histories, their works have some limitations. They outline historical pollution as the sole driver for establishing or amending environmental laws, while in reality, various factors can trigger legislative amendments. Furthermore, these studies focus primarily on a few nations, including the United States, the European Union, Australia, Germany, Canada and South Africa, while overlooking Latin America and West or East Africa. In the latter regions, mining practices differ significantly and are heavily influenced by artisanal and small-scale mining (ASM). Additionally, the roles of international organizations and community engagement are not well addressed in the said studies.

This book, highlights policy statements on environmental protection and community engagement and delves into laws that build on those areas. Regulations that explain how the corresponding laws shall be implemented are also highlighted.

Acid Mine Drainage: Sustainable Mining Practices for Mitigating Environmental Impact
DOI: https://doi.org/10.1016/B978-0-443-45107-2.00001-0

6.1 Evolution and characteristics of legislative frameworks for AMD

The evolution of legislative frameworks for AMD reflects a historical shift from unregulated resource extraction to a more refined system of environmental accountability. Early mining laws often focused solely on resource extraction but failed to consider ecological impacts, leaving a toxic legacy of abandoned mines as a public burden. Recognizing this gap, governments worldwide gradually moved toward integrated legislation aimed at reducing long-term environmental damage. As a result, modern environmental laws related to AMD are characterized by (1) multisectoral management that coordinates actions across water, mining and environmental agencies, (2) the institutionalization of the "polluter pays" principle, which ensures reclamation is funded through industry levies rather than public taxes, and (3) their inherently dynamic nature.

6.1.1 Multisectoral management

The AMD regulations are managed through a multisectoral approach, as no single act addresses them exclusively, requiring a combination of various legislations and government actors. Often, the legislations and regulatory bodies responsible for mining, water resources and environmental protection incorporate AMD management into their mandates (Simate, 2021). For example, in South Africa, a right to a safe environment is defined in Section 24 of their 1996 Constitution, which guarantees everyone an environment that is not harmful to their health or well-being (Naidoo, 2017; Simate, 2021). This constitutional right is further supported by the National Environmental Management Act of 1998 and the National Water Act (NWA) of 1998, which provide the legal framework for protecting water resources from pollution and ensuring their sustainable use (Humby, 2013; Kengni & Mostert, 2022).

In the United States, Federal laws set national standards, while individual states can implement additional regulations and enforcement mechanisms through state-level agencies. The US Federal laws include the National Environmental Policy Act (NEPA) of 1969. NEPA requires Federal agencies to prepare environmental impact statements for major Federal actions that may significantly affect the environment to make them available to public officials and citizens before actions are taken (Jessee, 1998). The Surface Mining Control and Reclamation Act (SMCRA) of 1977 has the main functions of regulating environmental impact of active coal mines and reclamation of abandoned mine land (AML). The Federal Water Pollution

Control Act, commonly referred to as the Clean Water Act (CWA) of 1977, requires mining operations to meet standards for surface water quality and for controlling discharges to surface water. The CWA-based regulations cover such mining-related situations as the disposal of mining-related waters, the pumping or draining of mine water to the surface, storm water runoff in mining operation areas and control of seeps from mine tailings impoundments (Barnhisel et al., 2015). It is from these Federal laws that states have an option to adopt regulations for environmental protection or establish their own. In California, for example, the California Toxic Rule (CTR) establishes numeric water quality criteria for toxic pollutants in California's inland surface waters, enclosed bays and estuaries to protect human health and aquatic life from priority pollutants (Sharifi et al., 2011). The priority pollutants associated with AMD are As, Cd, Cu, Pb, Ni and Zn (Jacobs & Testa, 2014). California addresses abandoned and inactive mine sites through the Surface Mining and Reclamation Act of 1975, which is overseen by the California State Mining and Geology Board (Kondolf, 1994).

6.1.2 Polluter pays and levies

Environmental regulations worldwide have become increasingly strict, with the "polluter pays principle" now firmly embedded in many legal systems (Gaur et al., 2022; Wiesmeth, 2021). This principle requires those responsible for pollution to bear the costs of preventing, mitigating, or remediating environmental damage (Moeng, 2019; Obreque-Contreras et al., 2015). It is implemented through pollution taxes, license fees and other financial tools designed to discourage harmful practices (Hansen et al., 2010). Polluter–pays taxes differ from general environmental taxes because they specifically target the liable polluter. Additional enforcement methods include compensatory damages and environmental reparation requirements within pollution control laws. Higher pollution taxes motivate polluters to adopt cleaner technologies, promote economic efficiency and ensure fairness by shifting cleanup costs from the public to the polluter (Garske et al., 2025; Ngosso, 2026).

6.1.3 Dynamic nature of environmental legislation for AMD

Environmental legislation is updated from time to time. Specific factors regarding AMD that led to changes in environmental regulations are (1) historical pollution, (2) advancements in science and technology and (3) international interventions (Mogashane et al., 2025; Spirnak et al., 2025).

6.1.3.1 *Historical pollution*

The history of environmental legislation has contributed to the transformation of environmental regulations. For instance, in the United States, toxic orange is on the Susquehanna River and other Appalachian waterways (Cravotta et al., 2010; Kruse Daniels et al., 2021; Merovich et al., 2021). Following visible fish kills and the turning of rivers orange in places like Pennsylvania (USA) (Cravotta et al., 2010; Mafane et al., 2025) and Mpumalanga (South Africa), governments moved to enforce strict, mandatory treatment regulations for active mines (Mhlongo & Amponsah-Dacosta, 2016). For South Africa, these events led to the strengthening of the National Water Act, which holds mining companies liable for post-closure water management. In the United States, the passage of the Federal Law on SMCRA of 1977 created the AML fund, which collects fees from active coal mining operations to pay for the cleanup of abandoned mines (Jacobs & Testa, 2014; Simate, 2021).

6.1.3.2 *Advancements in science and technology*

In the context of AMD, scientific and technological advancements act as a two-way traffic force that shapes mining laws and policies. Increasing industrialization and the rise of modern electronics and defense systems have surged demand for energy and critical minerals, leading to resource depletion and environmental stress. This surge induces governments to tighten mineral extraction regulations to address security and sustainability risks. Conversely, innovation provides the solutions to these challenges. Further, the development of greener extraction methods and advanced detection tools allows for the discovery of hidden pollutants and more effective AMD management. Thus, advancement in technology simultaneously accelerates mineral depletion and provides the tools for sustainable reclamation. Modern legislation and policies must therefore advocate for transformation to mandate utilization of higher technical standards (Al-Makhzoumi & AbuAbdoun, 2026). This ensures that industrial growth is balanced against long-term ecological health and the interests of local communities and social well-being in general.

China is significantly tightening its mining policies to prioritize environmental protection, shifting away from high-pollution development toward green mining and ecological restoration, aligned with the 2030 carbon peak and 2060 carbon neutrality targets (Hepburn et al., 2021). For instance, according to the Global Environmental Performance Index Report, China now ranks 120th among the world's major countries in overall environmental performance (Wang et al., 2023). In 2024 China revised its Mineral Resources Law for the third time since it came into effect in 1986, with previous revisions in 1996 and 2009 to add a chapter on "Ecological Restoration of Mining Areas." A key legal characteristic of the law regarding environmental issues is that mining and environmental governance in China are typically under state ownership (Wang et al., 2023).

6.1.3.3 International interventions

Global regulatory principles governing mining and AMD management are shaped by a range of international organizations, financial institutions and intergovernmental bodies. Although these global standards do not supersede host–country legislation, they provide influential benchmarks that guide responsible mining practices worldwide. Key contributors to these standards include the World Bank (World Bank, 2024); the International Finance Corporation (IFC); the World Health Organization (WHO); Intergovernmental Forum on Mining, Minerals, Metals and Sustainable Development (IGF); and the United Nations Environment Programme (UNEP) (UNEP, 2020). There are also nongovernmental organizations, such as International Institute for Sustainable Development (Baršauskaitė et al., 2025), International Network for Acid Prevention (INAP, 2009), International Mine Water Association (IMWA, 2026), Principles for Responsible Investment (PRI) (PRI, 2026) and International Council on Mining and Metals (ICMM) (ICMM, 2025), to mention a few. International organizations have been instrumental in promoting global sustainability principles in mining governance (IGF, 2026). The IGF, for example, provides guidance to governments on environmental management, mining sector governance and the application of sustainability standards (IGF, 2026). This guidance emphasizes the need to address sensitive issues such as AMD to prevent adverse environmental and social impacts while also discouraging corporate misconduct arising from noncompliance with sustainability principles (IGF, 2021). The UNEP, PRI and ICMM coconvened the Global Tailings Review in 2019 to establish an international standard for the safer management of tailings facilities (ICMM, 2019). The resulting standard, the Global Industry Standard on Tailings Management (GISTM), was launched on August 5, 2020 (Morrison, 2021). The GISTM strengthens environmental protection, enhances community transparency and requires rigorous technical, social and governance practices across a facility's entire lifecycle standards (Howe et al., 2025; Samadi et al., 2025).

The ICMM has developed a set of principles grounded in sustainable mining practices (Phillips, 2001). These principles cut across environmental, social and governance dimensions. Under the environmental performance category, ICMM's performance expectations require mining companies to design systems that prevent environmental harm, implement mitigation measures to avoid catastrophic failures and ensure pollution prevention. Principle 7, which focuses on biodiversity conservation, prohibits exploration or development of new mines within UNESCO World Heritage Sites and requires assessment of material risks that may lead to biodiversity loss. Collectively, these principles safeguard ecosystems and help prevent major impacts associated with AMD (ICMM, 2024).

According to INAP (2009), the IFC Environmental, Health and Safety Guidelines for Mining outlines several key parameters relevant to AMD management—five key ones include (1) design, operation and maintenance of TSFs in accordance with internationally recognized standards and risk–based strategies; (2) geochemical characterization of ore and waste, including leaching tests, ARD mapping and implementation of preventive measures; (3) rigorous environmental impact assessment conducted by qualified professionals; (4) mine closure and postclosure planning integrated into project design; and (5) financial feasibility and closure assurance that ensure long–term environmental protection. The WHO standards further define acceptable exposure levels for human health, providing thresholds for determining when drainage becomes harmful (WHO, 2013). Other responsibe mining frameworks include the ICMM Principles (ICMM, 2003), the International Cyanide Management Code (ICMI, 2021), the Kimberley Process Certification Scheme (KPCS), the British Columbia Guidelines for Metal Leaching and ARD , the Minerals Council of Australia's Enduring Value Framework, and the Mining Association of Canada's Towards Sustainable Mining Program (INAP, 2009).

Environmental legislations are widely used to regulate AMD, typically combining laws, regulations and practice guidelines. Across jurisdictions, several common regulatory themes emerge: (1) mandatory environmental impact assessments before authorization or permitting; (2) environmental management plans detailing how impacts will be managed during operations; (3) prediction and planning for mine closure, including AMD considerations; (4) ongoing monitoring, review and continuous improvement in AMD management; and (5) compliance with commitments specified in permits or EIA approvals (INAP, 2009; Verburg et al., 2009).

6.1.4 Sustainable development goals

Sustainable development can be defined as macroeconomic growth that enhances human well–being while remaining within the ecological carrying capacity of the environment (Cook & Davíðsdóttir, 2021; Monteiro et al., 2021; Tang & Qin, 2024; Wang et al., 2018; Zhang et al., 2022). A growing commitment to sustainable development has become a defining trend in shaping mining policies and laws in many countries (Bogers et al., 2022; van Driel et al., 2024). The global acceptance of sustainable development principles stems from the recognition that development must be socially, environmentally and economically sustainable (Ghimire, 2023; Murphy, 2012). A sustainable economy emerges from sustainable development practices that prioritize the preservation of natural resource bases through improved knowledge, organizational capacity, technical efficiency and long–term environmental stewardship (Glazewski & Posnik, 2000). Across the world, jurisdictions that adopt sustainable development practices have significantly influenced environmental laws in their nations, with direct implications for the mining industry, affecting everything from exploration and mine development to processing and mine closure (Monteiro et al., 2021; Tang & Qin, 2024).

In 2015 all United Nations member states adopted Agenda 2030 for sustainable development, which provides a shared blueprint for peace and prosperity for people and the planet, now and in the future. This comprises 17 Sustainable Development Goals (SDGs): (1) No Poverty, (2) Zero Hunger, (3) Good Health and Well-being, (4) Quality Education, (5) Gender Equality, (6) Clean Water and Sanitation, (7)Affordable and Clean Energy, (8) Decent Work and Economic Growth, (9) Industry, Innovation and Infrastructure, (10) Reduced Inequality, (11) Sustainable Cities and Communities, (12) Responsible Consumption and Production, (13) Climate Action, (14) Life Below Water, (15) Life on Land, (16) Peace, Justice and Strong Institutions, and (17) Partnerships for the Goals. These 17 SDGs are grouped into the three pillars of sustainable development: environmental, social, and economic. This translates to maximizing economic efficiency by repurposing waste into new materials, protecting the environment through water recycling and land reclamation and securing social equity by ensuring mineral extraction creates lasting infrastructure and fair labor for local communities.

6.2 Role of stakeholders in AMD management

Effective management of AMD requires coordinated action among multiple stakeholders. Internationally, the ICMM established sustainability principles that emphasize transparent and proactive stakeholder engagement, respect for human rights and inclusion of vulnerable groups, independent verification of environmental and social performance, and public reporting on sustainability indicators (ICMM, 2025). These global standards influence national policy frameworks and corporate practices in any country. Mining companies are the primary duty–bearers in AMD management. Their responsibilities include conducting environmental assessments, developing AMD management and mitigation plans, implementing monitoring systems, reporting on environmental performance, and engaging with communities to earn a "social license to operate" (Mills et al., 2025). The concept of a social license refers to broad community acceptance of mining operations, based on trust, transparency and demonstrated environmental responsibility (Verburg et al., 2009).

6.2.1 Consultation and communication

Effective consultation and communication are central to the management of AMD. Among the primary objectives of stakeholder consultation are evaluating the accuracy of predictions about future drainage quality, assessing the effectiveness of proposed mitigation measures, identifying uncertainties and developing contingency strategies to address them. These processes ensure that AMD risks are understood, anticipated and managed responsibly. Stakeholders' perceptions of AMD management significantly influence a company's social license to operate (Baloyi et al., 2024). Even where the risk of AMD is low, mining companies must demonstrate transparency and preparedness. Open communication builds trust and credibility, enhances stakeholder understanding of good mining practice, reduces misinformation and rumors and fosters a supportive environment for informed decision–making (Cruickshank et al., 2026). In this sense, communication is not merely a procedural requirement—it is a core component of responsible business practice. Within mining companies, communication functions such as public affairs, public relations, or community relations must collaborate closely with technical departments (Kemp, 2010; Kunz et al., 2017). Skilled personnel are often required to explain the scientific aspects of AMD in ways that are accessible to nontechnical audiences.

6.3 Policy, governance and community engagement for artisanal and small-scale mining (ASM)

The ASM has played an increasingly active role in improving people's livelihoods worldwide. As of 2024 the ASM employs 225 million people, both directly and indirectly; of these, 45 million are directly employed across Latin America, South America, Africa and Asia. ASM deals with several minerals ranging from well-known minerals and metals, such as gold, cobalt, copper and semiprecious and precious gemstones, to lesser known materials, such as salt, gravel and quarry rock (Boafo & Arthur-Holmes, 2025; Cartier, 2019; Kaaya et al., 2026; Ondayo et al., 2023). The ASM is therefore considered an important sector for reducing poverty. The ASM sector is receiving support from international organizations such as the UNEP and the Minamata Convention on Mercury (Minamata Convention on Mercury, 2024). Transforming ASM to the formal sector has been a challenge for decades (Table 6.1). In sub-Saharan Africa, for instance, there is significant variation in the level of formalization of ASM. Some countries have advanced formalization policies, while others are in the early stages of policy reform, as reported by Donkor et al. (2025) (Table 6.1). However, as of 2025, the available policies are unable to address environmental degradation and health risks caused by ASM activities in sub-Saharan African countries due to limitations such as weak enforcement, corruption, a high level of informality, insufficient community engagement, and limited regional coordination (Donkor et al., 2025).

According to the World Bank (Word Bank, 2024), the strategy of *Achieving Sustainable and Inclusive Artisanal and Small-Scale Mining (ASM): A Renewed Framework for World Bank Engagement* is built around nine core intervention areas divided into two main pillars, namely, Legitimacy and Professionalization. The nine interventions are (1) ensure responsive and appropriate regulatory frameworks, (2) expand application of progressive mine site standards, (3) enable gender equality across the value chain, (4) support appropriate decentralization and local governance models, (5) advance environmental stewardship and security, (6) facilitate financial inclusion, (7) support infrastructure to access legal markets, (8) build organizational capacity of ASM actors and (9) deliver extension services for miners. Word Bank (2024) reviewed each of the nine intervention areas and came up with a new proposal of supporting government ASM actors and mining communities. Regarding environmental stewardship, World Bank intends to build off expertise in mine closure and rehabilitation in the large-scale mining sector along with tested methodologies for rehabilitation in the ASM sector to help governments and communities manage legacy and liability mines (Word Bank, 2024).

Table 6.1 A comparative table, detailing ASM-related policies in some selected sub-Saharan Africa countries (Grouped by Common Policy Experiences). From Donkor et al. (2025).

Country	Policy Focus	Key Policy Framework Highlights	Strengths	Weaknesses	Implementation Status
Group A: Countries with Advanced and Progressive Formalization Efforts					
Rwanda	Promoting formalization, sustainability	2018 Mining Law; international partnerships for cooperative mining	Strong governance; cooperative model and transparency	Informality persists; weak enforcement	Progressive, with growing success
Botswana	Formalization, economic empowerment and sustainability	Mining policy for long-term growth; inclusive mining policies	Strong institutions; high economic contributions;	Limited ASM representation	ongoing process
Ghana	Formalization and sustainable mining	Act 703 of 2006 (Amended in 2019, see e. g., Act 995); licensing, cooperatives	Encourages formalization; accessible	Weak enforcement; lack of political will; corruption	Partial
Guinea	Formalization and integration	2011 Mining Code; simplified licensing	Strong framework; state commitment	Informality and weak enforcement	Partial
Group B: Countries with Partial Formalization and Strong Environmental Focus					
Tanzania	Formalization, environmental compliance	Environmental Management Act	Strong environmental laws	High costs; low compliance	Partial
Mozambique	Environmental management & formalization	ASM zoning and mercury reduction programs	Strong international collaboration	Informality; fragmented progress	In progress
Senegal	ASM licensing & environmental sustainability	Mining cooperatives and environmental rules	Awareness of sustainability	Monitoring challenges	Partial
Uganda	Regulation & health/ environment focus	2003 Mining Act; gender-inclusive reforms	Government engagement	weak Enforcement; informality high	Partial
South Africa	Economic Health & Safety	Mine empowerment and Act; Broad-Based Black Safety Economic Empowerment (BBBEE)	Worker protections	High illegality; costly compliance	Gradual

(continued on next page)

Group C: Countries with Post-Conflict Reforms and Fragile Governance					
Sierra Leone	Post-conflict sector reform	Diamond certification; legal mining	Reduces exploitation	Corruption in enforcement	Progressing steadily
South Sudan	Economic recovery via ASM	2012 Mining Act; international collaboration	Recognized role of ASM	Instability; informality	Limited
Central African Republic	Conflict mineral regulation	ASM formalization to reduce conflict funding	ASM supports livelihoods	Linked to conflict; weak governance	Limited
Liberia	Formalization and environmental protection	Internationally supported reforms	Local benefits	Weak enforcement	Ongoing
DR Congo	Formalization & Conflict mineral management	Gold and cobalt-focused formalization	Economic relevance	Conflict, child labor issues	Slow due to instability
Group D: Countries with Market Access and Economic Development Focus					
Mali	Revenue generation and	Revenue generation and exports Tax simplification	export facilitation Encourages trade	Limited sustainability	Actively pursued
Zimbabwe	Formalization and market access	Gold Service Centers, registration	Promotes legal market entry	Corruption; informality	Limited success
Zambia	Diversification and safety	Mines and Minerals Development Act	Economic diversification	Weak community engagement	Partial
Namibia	Export optimization	National Mining Policy	GDP contribution	Lacks ASM focus	Progressing steadily
Nigeria	Formalization and environmental protection	ASM policy within mining laws	Awareness of ASM's role	Weak monitoring; informality	Partial

(*continued on next page*)

Table 6.1 A comparative table, detailing ASM-related policies in some selected sub-Saharan Africa countries (Grouped by Common Policy Experiences). From Donkor et al. (2025)—cont'd

Group A: Countries with Advanced and Progressive Formalization Efforts

Country	Policy Focus	Key Policy Framework Highlights	Strengths	Weaknesses	Implementation Status
Group E: Countries in Early Stages of ASM Policy Reform					
Ethiopia	Formalization and poverty alleviation	Cooperative mining promotion	Community development	Logistics, enforcement	Early stage
Burundi	Formalization and integration	2013 Mining Code	Livelihood recognition	Institutional weakness	Early stage
Malawi	Economic development and formalization	Mines & Minerals Act	Employment generation	Limited technical support	Partial
Eritrea	Revenue-focused formalization	Tax-based regulation	Revenue potential	High informality	Limited
Madagascar	Formalization and environmental policy	Permit-based ASM regulation	Environmental attention growing	Widespread informality	Partial
Burkina Faso	Formalization and equitable mining	2015 Code; decentralized licensing	Strong state commitment	Persistent informality, limited enforcement	Uneven
Chad	ASM registration and poverty reduction	Licensing policy; technical support lacking	Economic potential	Informality dominates	Minimal
Group F: Countries with Emerging or Nascent Policy Frameworks					
Angola	Diamond-focused ASM regulation	Mining Code 2011; Endiama oversight	Government interest	Informality; low transparency	Partial
Cote dˆ Ivoire	Formalization and local development	2014 Mining Code; decentralized control	Strong legal base	Institutional friction	Partial
Niger	Formalization, Licensing and environmental control	2006 Mining Code emphasizes ASM registration, licensing, and environmental compliance	Promotes legal ASM	Weak support for miners	Limited
Kenya	Environmental conservation and licensing	2016 Mining Act; inclusive approach	Comprehensive framework	Weak monitoring	Slow progress

The evolution of legislative frameworks for AMD signifies a global shift from prioritizing raw resource extraction to enforcing rigorous environmental accountability. The institutionalization of the Polluter Pays Principle has fundamentally changed the economic landscape of mining, shifting the financial burden of reclamation from the public to industry. However, as this chapter has highlighted, the dynamic nature of these laws must continually adapt to lessons from historical pollution, international standards and technological advances. While developed nations have established robust benchmarks, the future of AMD policy lies in addressing the unique challenges of ASM and deepening community engagement to ensure that legal frameworks are not only technically sound but also socially justifiable.

6.4 Policies and legislation in selected countries

This section evaluates legal and regulatory frameworks governing AMD management across selected jurisdictions, including Canada, Australia, the European Union (EU), Indonesia and Tanzania. Comparative assessment of these systems reveals substantial differences in the maturity, technical specificity and implementation capacity of AMD governance between developed and developing countries. The review further identifies regulatory best practices and critical gaps that influence the effectiveness of AMD prediction, prevention, mitigation and long-term environmental management under diverse socio-economic conditions.

6.4.1 Canada

Canada's federal governance system, consisting of ten provinces and three territories, creates a shared distribution of legislative responsibilities for environmental and mining management (Jackson et al., 2023). Consequently, AMD management is governed through an integrated framework involving federal regulations, provincial legislation and technical guidance documents. Canadian legislation relevant to AMD encompasses environmental impact assessment requirements, acid rock drainage (ARD) guidance documents, mine closure regulations, financial assurance provisions and effluent discharge standards. Major legislative instruments and technical guidance identified by INAP (2009) include: (1) *Guidelines for Metal Leaching and Acid Rock Drainage at Mine Sites in British Columbia* (Price & Errington, 1998); (2) *Guidelines for ARD Prediction in the North*; (3) *Metal Mining Effluent Regulations*

(MMER); (4) provincial water-quality discharge standards; and (5) mine closure and financial assurance requirements. Mine closure planning under Canada's *Minerals and Metals Policy* requires detailed assessment of waste-rock management, tailings disposal systems, drainage controls and long-term environmental performance (Shinya, 1998). Particular emphasis is placed on demonstrating long-term physical and geochemical stability through evaluations of acid generation potential, metal leaching behavior, water-quality impacts and post-closure treatment requirements. This comprehensive risk-based approach allows environmental liabilities to be identified and addressed during early stages of mine planning rather than after mine closure.

6.4.2 Australia

Australia has established a comprehensive regulatory framework for AMD management designed to protect environmental assets, including biodiversity, water resources and landscape integrity (Mudd, 2007). Regulatory implementation largely occurs through environmental approval processes requiring mining proponents to demonstrate effective mitigation of environmental impacts throughout mine development, operation and closure. Key considerations within Australian AMD-related legislation include: (1) identification and assessment of AMD risks during environmental and social impact assessments; (2) determination of financial assurance mechanisms to secure adequate resources for post-closure environmental management; (3) compliance with national water-quality standards, including monitoring and reporting obligations; and (4) assessment of local and regional water-resource availability and vulnerability to contamination. Technical guidance is provided through the *Leading Practice Sustainable Development Program for the Mining Industry: Managing Acid and Metalliferous Drainage* (Jones et al., 2016). Unlike broad legislative provisions, this document provides practical methodologies for prediction, prevention and management of AMD, thereby strengthening consistency between regulatory expectations and operational implementation.

6.4.3 European Union

Within the European Union (EU), the European Commission is responsible for proposing legislation that is subsequently adopted by the European Parliament and the Council of the European Union (European Parliament, 2026). While the Commission establishes minimum regulatory requirements, member states retain the flexibility to adopt stricter environmental standards through national legislation. The AMD management within the EU is supported through multiple interconnected environmental directives. Among these, the *Directive on the Management of Waste from Extractive Industries* (2006/21/EC), commonly referred to as the *Mine Waste Directive*, represents one of the principal legislative instruments governing mining waste management (Karamfilova, 2017). Other directives relevant to AMD management include: *Environmental Impact Assessment Directive* (European Commission: Directorate-General for Environment, Milieu, COWI, 2025), which requires assessment of direct and indirect environmental impacts of proposed mining projects; *Water Framework Directive* (European Commission: Directorate-General for Environment, Valerie Fogleman, 2025), which establishes procedures for water-body classification and river-basin management; *Environmental Liability Directive* (>European Commission: Directorate-General for Environment, 2025), which establishes responsibilities for remediation of environmental damage and associated costs; and *Major Accidents Involving Dangerous Substances Directive* (2003/105/EC), derived from the Seveso II framework, which focuses on prevention and mitigation of industrial accidents involving hazardous substances (Wood, 2009). Collectively, these directives provide a structured and integrated regulatory system for AMD management, while allowing individual member states flexibility to implement stricter environmental requirements.

6.4.4 Indonesia

In Indonesia, AMD and mining wastewater management are principally regulated through the *Decree of the State Minister of Environment No. 113 of 2003*, which establishes wastewater quality standards for coal-mining and coal-processing activities (Afaz et al., 2025; Pratiwi et al., 2021). The regulation establishes permissible discharge limits for critical water-quality parameters intended to protect human health and environmental quality. These include pH, total suspended solids (TSS), iron, manganese, chemical oxygen demand (COD), biochemical oxygen demand (BOD_5), oil and grease, chromium, cadmium, copper, zinc, lead, cyanide and phenolic compounds.The regulation further requires mining operators to conduct routine environmental monitoring, maintain compliance records and implement treatment systems capable of maintaining effluent quality within prescribed thresholds before discharge into receiving environments (Ekariyono & Sugiarto, 1994). Such quantitative standards provide measurable benchmarks for environmental compliance and facilitate regulatory monitoring.

6.4.5 Tanzania

Environmental protection and sustainable management of mining activities are critical in Tanzania because of the country's diverse mineral resources, ecological systems and transboundary water networks (NEMC, 2024). Tanzania possesses substantial mineral wealth, including gold, diamonds, coal and the globally unique gemstone tanzanite, alongside emerging strategic resources such as nickel, graphite, rare-earth elements and lithium. Sustainable exploitation of these resources requires regulatory mechanisms capable of minimizing environmental degradation while safeguarding natural ecosystems and water resources for future generations. The *National Environmental Policy (NEP) of 1997* remains the foundation of environmental governance in Tanzania (Vice President's Office, 2021). The policy identifies major environmental concerns, including land degradation, water-resource deterioration, environmental pollution, biodiversity loss, degradation of aquatic systems and deforestation. To implement the policy, the Government enacted the *Environmental Management Act (EMA) of 2004*, which established a legal and institutional framework for sustainable environmental management. EMA provides overarching environmental governance requirements applicable to all development sectors, including mining. Supporting regulations include the *Environmental Management (Environmental Impact Assessment and Audit) Regulations of 2005 and the Environmental Management (Registration and Practice of Environmental Experts) Regulations of 2021.* Environmental Impact Assessment (EIA) is mandatory for mining operations requiring Special Mining Licenses (SMLs) and Mining Licenses (MLs). These assessments are intended to identify potential environmental impacts and define appropriate mitigation measures. In addition, the *Mineral Policy of 2009 and the Mining Act of 2010* establish environmental responsibilities for mining operations. Key provisions include strengthening institutional capacity for environmental monitoring and enforcement, requiring financial provisions for mine closure and rehabilitation, harmonizing environmental legislation and promoting environmental awareness among mining communities. Mine closure planning is further guided by the *Mine Closure Guidelines of 2009*, developed in accordance with the Environmental Management Act, Water Resources Management Act and Mining Act. The guidelines require consideration of long-term environmental impacts associated with mining activities, including waste management, land rehabilitation and water protection. Small-scale mining operations operating under Primary Mining Licences (PMLs) are required to prepare Environmental Protection Plans (EPPs) (NEMC, 2018). These plans generally include descriptions of mineral deposits, mining methods, infrastructure, waste management procedures, chemical use and potential environmental emissions.Although Tanzania's regulatory framework demonstrates a strong commitment to environmental protection and sustainable mining practices, specific technical guidelines addressing AMD prediction, characterization and mitigation remain limited. Consequently, implementation may depend heavily on individual project assessments and practitioner expertise, potentially leading to inconsistencies in AMD risk evaluation.

6.4.6 Conclusion and recommendations

The comparative assessment demonstrates considerable variation in AMD governance among the selected jurisdictions. Countries with long mining histories, including Canada, Australia and EU member states, generally employ mature regulatory systems supported by detailed technical guidelines for AMD prediction, prevention, monitoring and long-term management. These systems commonly integrate risk-based assessments, financial assurance mechanisms and post-closure environmental obligations. In contrast, developing countries frequently rely on broader environmental legislation without dedicated AMD-specific procedures and technical standards. Tanzania illustrates this challenge, where environmental legislation provides a robust legal framework but lacks standardized protocols for AMD characterization and prediction.Studies conducted in Tanzania (Almås & Manoko, 2012; Kaaya et al., 2026) suggest that limited baseline geochemical characterization and absence of standardized assessment methodologies remain important challenges. Additional constraints may include limited analytical capacity, insufficient technical expertise and restricted regulatory monitoring capabilities.Future policy development should therefore prioritize: establishment of national AMD-specific guidelines for prediction and management; development of standardized geochemical characterization protocols for mine waste materials; strengthening laboratory and technical capacity for AMD investigations; enhancement of environmental monitoring and regulatory enforcement systems; and increased collaboration among government agencies, academia and industry stakeholders. Implementation of these measures would improve scientific consistency, strengthen regulatory effectiveness and enhance long-term environmental sustainability within mining regions.

References

Afaz, T., Gusman, M., Barlian, E., & Syah, N. (2025). Analysis of the Impact of Coal Mine Acid Water on River Water Quality and Its Implications for River Water User Communities. *Jurnal Penelitian Pendidikan IPA, 11*(2), 333–341. https://doi.org/10.29303/jppipa.v11i2.9645.

Al-Makhzoumi, O., & AbuAbdoun, M. M. M. (2026). From government policies to digital transformations: A modern perspective on environmental legislation and its impact on sustainable development. In *Artificial intelligence for sustainable innovation managemententent and risk management* (pp. 1171–1183). Springer. https://doi.org/10.1007/978-3-031-95310-1_86.

Almås, Å. R., & Manoko, M. L. K. (2012). Trace Element Concentrations in Soil, Sediments, and Waters in the Vicinity of Geita Gold Mines and North Mara Gold Mines in Northwest Tanzania. *Soil and Sediment Contamination, 21*(2), 135–159. https://doi.org/10.1080/15320383.2012.649372.

Baloyi, J., Ramdhani, N., Mbhele, R., & Simate, G. S. (2024). Acid mine drainage from gold mining in South Africa: Remediation, reuse, and resource recovery. *Mine Water and the Environment, 43*(3), 418–430. https://doi.org/10.1007/s10230-024-00994-2.

Barnhisel, R. I., Darmody, R. G., & Daniels, W. L. (2015). Coal mine reclamation, acid mine drainage, and the clean water act. In *Reclamation of drastically disturbed lands: 41* (pp. 169–191). Wiley. https://doi.org/10.2134/agronmonogr41.c7.

Baršauskaitė, I., Gubler, J., Moerenhout, T., Nikiema, S., Ostřanský, J., Tipping, A., & Verma, R. (2025). *International trade and investment agreements and sustainable critical minerals supply: IISD report*. International Institute for Sustainable Development (IISD). https://www.iisd.org/.

Boafo, J., & Arthur-Holmes, F. (2025). Sustainability transition paradox: Emerging dimensions of illegal artisanal and small-scale mining of critical minerals in Africa. *Resources Policy, 108*, 105673. https://doi.org/10.1016/j.resourpol.2025.105673.

Bogers, M., Biermann, F., Kalfagianni, A., & Kim, R. E. (2022). Sustainable development goals fail to advance policy integration: A large-n text analysis of 159 international organizations. *Environmental Science and Policy, 138*, 134–145. https://doi.org/10.1016/j.envsci.2022.10.002.

Cartier, L. E. (2019). Gemstones and sustainable development: Perspectives and trends in mining, processing and trade of precious stones. *Extractive Industries and Society, 6*(4), 1013–1016. https://doi.org/10.1016/j.exis.2019.09.005.

Cheang, B. (2026). The incoherence of modest industrial policy. *International Studies in Entrepreneurship, 49*, 61–80. https://doi.org/10.1007/978-3-032-15512-2_3.

Christopher Ulua, K., & Uju, R. E. (2021). Industrial policy and sustainable development: Theoretical perspectives. In *Industry, innovation and infrastructure* (pp. 1–14). Springer International Publishing. https://doi.org/10.1007/978-3-319-71059-4_141-1.

Cook, D., & Davíðsdóttir, B. (2021). An appraisal of interlinkages between macro-economic indicators of economic well-being and the sustainable development goals. *Ecological Economics, 184*, 106996. https://doi.org/10.1016/j.ecolecon.2021.106996.

Cravotta, C. A., III, Brightbill, R. A., & Langland, M. J. (2010). Abandoned mine drainage in the Swatara Creek Basin, Southern Anthracite Coalfield, Pennsylvania, USA: 1. Stream water quality trends coinciding with the return of fish. *Mine Water and the Environment, 29*(3), 176–199. https://doi.org/10.1007/s10230-010-0112-6.

Cruickshank, S., McKee, M., & Pagel, C. (2026). Effective communication and public engagement strategies to counter misinformation about infectious diseases. *Immunology & Cell Biology, 104*(2), 92–105. https://doi.org/10.1111/imcb.70073.

Deberdt, R., & Letourneau, A. (2025). Establishing a critical minerals industry: Review of Canada's administrative and legal efforts at the national and sub-national levels. *Resources Policy, 107*, 105660. https://doi.org/10.1016/j.resourpol.2025.105660.

Deberdt, R., Smith, N. M., Calderon, J. L., & McCall, S. K. (2025). Critical minerals lists for low-carbon transitions: Reviewing their structure, objectives, and limitations. *Energy Research and Social Science, 127*. https://doi.org/10.1016/j.erss.2025.104252.

DEMIRS. (2025). Guideline for preparing Mine Closure Plans. The Department of Energy, Mines, Industry Regulation and Safety (DEMIRS), Government of Western Australia, November. www.dmpe.wa.gov.au.

Donkor, P., Siabi, E. K., Mensah, S. K., Frimpong, K., Vuu, C., Nyantakyi, E. K., Siabi, E. S., & Ameyaw, A. B. (2025). Informal artisanal and small-scale mining in sub-saharan Africa: Policy evolution, sector formalization, and future trajectories. *Resources Policy, 109*(January). https://doi.org/10.1016/j.resourpol.2025.105698.

Ekariyono, W., & Sugiarto, S. (1994). Environmental Baseline and Impact Management of Jambu-1 Witaitonga Island - Southeast Sulawesi, Indonesia. In *SPE Health, Safety and Environment in Oil and Gas Exploration and Production Conference*. https://doi.org/10.2118/27120-ms.

European Commission: Directorate-General for Environment, Milieu, COWI, E. (2025). *Collection of information and data on the implementation of the revised Environmental Impact Assessment (EIA) Directive (2011 / 92 / EU) as amended by 2014/52/EU). Final Report*. Publications Office of the European Union. https://doi.org/10.2779/817391.

European Commission: Directorate-General for Environment, Valerie Fogleman, S. & B. L. (2025). *Facilitating enforcement of the Environmental Liability Directive (ELD) by competent authorities: Hungary*. Publications Office of the European Union. https://doi.org/10.2779/803841.

European Commission: Directorate-General for Environment. (2025). *Assessment of the third River Basin Management Plans under the Water Framework Directive Assessment of international cooperation under the Water Framework Directive – International River Basins*. Publications Office of the European Union. https://doi.org/10.2779/1983167.

European Partliament. (2026). *Legislative powers*. European Union. https://www.europarl.europa.eu/about-parliament/en/parliaments-powers/legislative-powers.

Garske, B., Heyl, K., & Ekardt, F. (2025). How economic instruments address sustainable nutrient use: The example of phosphorus governance. *Environmental Sciences Europe, 37*(1). https://doi.org/10.1186/s12302-024-01045-x.

Gaur, A., Gurjar, S. K., & Chaudhary, S. (2022). Circular system of resource recovery and reverse logistics approach: Key to zero waste and zero landfill. In *Advanced organic waste management: Sustainable practices and approaches* (pp. 365–381). Elsevier. https://doi.org/10.1016/B978-0-323-85792-5.00008-3.

Ghimire, B. J. (2023). Three pillars of sustainable development: Challenges versus achievements. *Journey for Sustainable Development and Peace Journal, 1*(2), 132–146. https://doi.org/10.3126/jsdpj.v1i02.58266.

Glazewski, J., & Posnik, S. (2000). Compliance with international environmental standards and expectations: Review of international developments. *Journal of the South African Institute of Mining and Metallurgy, 100*(4), 211–219.

Hansen, E., Collins, A., Zegre, S., & Hereford, A. (2010). The benefits of acid mine drainage remediation on the North Branch Potomac River. *Morgantown (WV). Downstream Strategies*.

Hepburn, C., Qi, Y., Stern, N., Ward, B., Xie, C., & Zenghelis, D. (2021). Towards carbon neutrality and China's 14th Five-Year Plan: Clean energy transition, sustainable urban development, and investment priorities. *Environmental Science and Ecotechnology, 8*, 100130. https://doi.org/10.1016/j.ese.2021.100130.

Howe, L., Cote, C., & Johnston, S. (2025). Applying high reliability organisation theory to guide implementation of the global industry standard for tailings management (GISTM). *Business Strategy and the Environment*. https://doi.org/10.1002/bse.70427.

Humby, T.-L. (2013). The spectre of perpetuity liability for treating acid water on South Africa's Goldfields: Decision in harmony II. *Journal of Energy & Natural Resources Law, 31*(4), 453–466. https://doi.org/10.1080/02646811.2013.11435343.

ICMI. (2021). The International Cyanide Management Code. International Cyanide Management Institute (ICMI). https://cyanidecode.org/about-the-cyanide-code/the-cyanide-code/.

ICMM. (2019). *ICMM, UN environment programme and principles for responsible investment agree to co-convene mine tailings storage facilities review*. https://www.icmm.com/en-gb/news/2019/tailings-review.

ICMM. (2025). Tool for acid rock drainage and metal leaching prevention and management. International Council on Mining and Metals (ICMM). https://www.icmm.com/en-gb/guidance/environmental-stewardship/2025/ardml.

ICMM. (2003). *Mining and protected areas*. International Council on Mining and Metals (ICMM).

ICMM. (2024). Mining Principles: Performance Expectations. In *International Council on Mining and Metals (ICMM)*. https://www.icmm.com/en-gb/about-us/member-requirements/mining-principles.

IGF. (2021). *Guidance for governments on environmental management, mining sector governance, and the application of sustainability standards*. https://www.igfmining.org/about-us/.

IGF. (2026). *The Intergovernmental Forum on Mining, Minerals, Metals and Sustainable Development (IGF)*. https://www.igfmining.org/about-us/.

IMWA. (2026). International Mine Water Association (IMWA). https://www.imwa.info/.

INAP. (2009). Global Acid Rock Drainage (GARD) guide. International Network for Acid Prevention. https://gardguide.com/.

Jackson, S., Poelzer, G., Poelzer, G., & Noble, B. (2023). Mining and Sustainability in the Circumpolar North: The Role of Government in Advancing Corporate Social Responsibility. *Environmental Management, 72*(1), 37–52. https://doi.org/10.1007/s00267-022-01680-1.

Jacobs, J. A., & Testa, S. M. (2014). Mine reclamation policy and regulations of selected jurisdictions. In J. A. Jacobs, J. H. Lehr, & S. M. Testa (Eds.), *Acid mine drainage, rock drainage, and acid sulfate soils* (pp. 319–324). Wiley. https://doi.org/10.1002/9781118749197.ch28.

Jessee, L. (1998). The National Environmental Policy Act net (NEP ANET) and DOE NEPA web: What they bring to environmental impact assessment. *Environmental Impact Assessment Review, 18*(1), 73–82. https://doi.org/10.1016/s0195-9255(97)00068-1.

Jones, D., Jeff, T., Sophie, P., McCullough, C., Brown, P., Garvie, A., … Robertson, A. (2016). Preventing acid and metalliferous drainage: Leading Practice Sustainable Development Program for the Mining Industry. Australia: Department of Industry, Science and Resources. https://www.industry.gov.au/publications/leading-practice-handbooks-sustainable-mining/preventing-acid-and-metalliferous-drainage.

Jonker-Hoffrén, P., & Sippola, M. (2024). Industrial policy. In *Encyclopedia of public policy* (pp. 1–6). Springer International Publishing. https://doi.org/10.1007/978-3-030-90434-0_107-1.

Kaaya, N. I., Vegi, M. R., & Macheyeki, A. S. (2026). Acid mine drainage, soil pollution, and carcinogenic risk of geogenic contaminants in artisanal and small-scale gold mining areas of Geita, Tanzania. *Environmental Geochemistry and Health, 48*(1), 23. https://doi.org/10.1007/s10653-025-02909-8.

Karamfilova, E. (2017). *Mining Waste Directive 2006/21/EC: European Implementation Assessment*. European Parliamentary Research Service.

Kemp, D. (2010). Community relations in the global mining industry: Exploring the internal dimensions of externally orientated work. *Corporate Social Responsibility and Environmental Management, 17*(1), 1–14. https://doi.org/10.1002/csr.195.

Kengni, B., & Mostert, H. (2022). Regulation and remediation of the impacts of mine wastes on land and water in South Africa. *Journal of Energy & Natural Resources Law, 40*(3), 337–360. https://doi.org/10.1080/02646811.2021.1985780.

Kondolf, G. M. (1994). Environmental planning in regulation and management of instream gravel mining in California. *Landscape and Urban Planning, 29*(2-3), 185–199. https://doi.org/10.1016/0169-2046(94)90027-2.

Kruse Daniels, N., LaBar, J. A., & McDonald, L. M. (2021). Acid mine drainage in Appalachia: Sources, legacy, and treatment. In *Appalachia's coal-mined landscapes* (pp. 193–216). Springer International Publishing. https://doi.org/10.1007/978-3-030-57780-3_8.

Kunz, N. C., Kastelle, T., & Moran, C. J. (2017). Social network analysis reveals that communication gaps may prevent effective water management in the mining sector. *Journal of Cleaner Production, 148*, 915–922. https://doi.org/10.1016/j.jclepro.2017.01.175.

Mafane, D., Ngulube, T., & Mabel Mphahlele-Makgwane, M. (2025). Anaerobic bioremediation of acid mine drainage using sulphate-reducing bacteria: Current status, challenges, and future directions. *Sustainability, 17*(8), 3567. https://doi.org/10.3390/su17083567.

Masindi, V., Foteinis, S., Renforth, P., Ndiritu, J., Maree, J. P., Tekere, M., & Chatzisymeon, E. (2022). Challenges and avenues for acid mine drainage treatment, beneficiation, and valorisation in circular economy: A review. *Ecological Engineering, 183*, 106740. https://doi.org/10.1016/j.ecoleng.2022.106740.

Merovich, G. T., Hitt, N. P., Merriam, E. R., & Jones, J. W. (2021). Response of aquatic life to coal mining in Appalachia. In *Appalachia's coal-mined landscapes* (pp. 245–285). Springer International Publishing. https://doi.org/10.1007/978-3-030-57780-3_10.

Mgala, E., & Gwambene, B. (2024). Reassessing water resource challenges in a changing climate in the Kagera transboundary river basin, Tanzania. *Physics and Chemistry of the Earth, Parts A/B/C, 136*, 103763. https://doi.org/10.1016/j.pce.2024.103763.

Mhlongo, S. E., & Amponsah-Dacosta, F. (2016). A review of problems and solutions of abandoned mines in South Africa. *International Journal of Mining, Reclamation and Environment, 30*(4), 279–294. https://doi.org/10.1080/17480930.2015.1044046.

Mills, L. N., Stewart, J. M., & Auld, G. (2025). Licensing to operate: Understanding variations in regulatory outcomes in the Australian mining sector. *Resources Policy, 105*. https://doi.org/10.1016/j.resourpol.2025.105578.

Minamata Convention on Mercury. (2024). Minamata Convention on Mercury: text and annexes. https://minamataconvention.org/en/resources/minamata-convention-mercury-text-and-annexes.

Minihane, M. R. (2012). Evaluation of streamflow estimates for the Rovuma River. *Physics and Chemistry of the Earth, 50–52*, 14–23. https://doi.org/10.1016/j.pce.2012.09.003.

Moeng, K. (2019). Community perceptions on the health risks of acid mine drainage: The environmental justice struggles of communities near mining fields. *Environment, Development and Sustainability, 21*(6), 2619–2640. https://doi.org/10.1007/s10668-018-0149-4.

Mogashane, T. M., Maree, J. P., Mokoena, L., & Tshilongo, J. (2025). Research activities on acid mine drainage treatment in South Africa (1998–2025): Trends, challenges, bibliometric analysis and future directions. *Water (Switzerland), 17*(15). https://doi.org/10.3390/w17152286.

Monteiro, N. B. R., Bezerra, A. K. L., Moita Neto, J. M., & Silva, E. A. D. (2021). Mining law: In search of sustainable mining. *Sustainability (Switzerland), 13*(2), 1–16. https://doi.org/10.3390/su13020867.

Morrison, K. F. (2021). Preface to the MME special focus issue on tailings management. *Mining, Metallurgy and Exploration, 38*(3), 1287–1288. https://doi.org/10.1007/s42461-021-00433-8.

Mteti, S. H., Mpambije, C. J., & Manyerere, D. J. (2025). Unlocking cultural tourism: Local community awareness and perceptions of cultural heritage resources in Katavi Region in southern circuit of Tanzania. *Social Sciences and Humanities Open, 11*(September 2024), 101295. https://doi.org/10.1016/j.ssaho.2025.101295.

Mudd, G. M. (2007). Gold mining in Australia: linking historical trends and environmental and resource sustainability. *Environmental Science & Policy, 10*(7–8), 629–644. https://doi.org/10.1016/j.envsci.2007.04.006.

Murphy, K. (2012). The social pillar of sustainable development: A literature review and framework for policy analysis. *Sustainability: Science, Practice, and Policy, 8*(1), 15–29. https://doi.org/10.1080/15487733.2012.11908081.

Naidoo, S. (2017). *Acid mine drainage in South Africa Development actors, policy impacts, and broader implications*. University of South Africa (UNISA).

Narbaitz, R. M., Akerele, G. O., Sartaj, M., & Downey, J. (2025). Chlorine regeneration of a zeolite ion-exchange column for ammonia removal from an explosives-impacted mining wastewater. *Environmental Technology, 46*(18), 3513–3526. https://doi.org/10.1080/09593330.2025.2471043.

NEMC. (2018). *Terms of Reference for Environmental Impact Assessment of Mining Project*. National Environment Management Council.

NEMC. (2024). *State of the coast for mainland Tanzania, Transitioning to Blue Economy: Contribution of Coastal and Marine Environment*. National Environment Management Council (NEMC) http://www.nemc.or.tz/.

Ngosso, T. (2026). Climate responsibilities of business corporations. *Interdisciplinary Reviews: Climate Change, 17*(1). https://doi.org/10.1002/wcc.70040.

Obreque-Contreras, J., Pérez-Flores, D., & Gutiérrez, P. (2015). Acid mine drainage in Chile: An opportunity to apply bioremediation technology. *Journal of Waste Water Treatment & Analysis, 6*(3). https://doi.org/10.4172/2157-7587.1000215.

Ondayo, M. A., Watts, M. J., Hamilton, E. M., Mitchell, C., Mankelow, J., & Osano, O. (2023). Artisanal gold mining in Kakamega and Vihiga counties, Kenya: Potential human exposure and health risk. *Environmental Geochemistry and Health, 45*(8), 6543–6565. https://doi.org/10.1007/s10653-023-01647-z.

Pellizzoni, L. (2004). Responsibility and environmental governance. *Environmental Politics, 13*(3), 541–565. https://doi.org/10.1080/0964401042000229034.

Phillips, A. (2001). *Mining and protected areas*. World Business Council for Sustainable Development and International Institute for environment and development, 62.

Pratiwi, Narendra, B. H., Siregar, C. A., Turjaman, M., Hidayat, A., Rachmat, H. H., Mulyanto, B., Suwardi, Iskandar, Maharani, R., Rayadin, Y., Prayudyaningsih, R., Yuwati, T. W., Prematuri, R., & Susilowati, A. (2021). Managing and reforesting degraded post-mining landscape in Indonesia: A review. *Land, 10*(6). https://doi.org/10.3390/land10060658.

PRI. (2026). *PRI home*. Responsible Investment (PRI). https://www.unpri.org/.

Price, W. A., & Errington, J. C. (1998). Guidelines for Metal Leaching and Acid Rock Drainage At Minesites in British Columbia. Ministry of Energy and Mines, August, 92.

Samadi, M., Gilani, S.-O., Abdollahisharif, J., & Bakhtavar, E. (2025). Towards sustainable mine closure and reclamation through strategic tailing disposal sites evaluation: A causality-based Dempster-Shafer framework. *Resources Policy, 108*, 105676. https://doi.org/10.1016/j.resourpol.2025.105676.

Seddon, D., Kashaigili, J. J., Taylor, R. G., Cuthbert, M. O., Mwihumbo, C., & MacDonald, A. M. (2021). Focused groundwater recharge in a tropical dryland: Empirical evidence from central, semi-arid Tanzania. *Journal of Hydrology: Regional Studies, 37*(February), 100919. https://doi.org/10.1016/j.ejrh.2021.100919.

Sharifi, S., Massoudieh, A., & Kayhanian, M. (2011). A stochastic stormwater quality volume-sizing method with first flush emphasis. *Water Environment Research, 83*(11), 2025–2035. https://doi.org/10.2175/106143011 × 12989211.

Sharygin, V. V., Kamenetsky, V. S., Zaitsev, A. N., & Kamenetsky, M. B. (2012). Silicate–natrocarbonatite liquid immiscibility in 1917 eruption combeite–wollastonite nephelinite, Oldoinyo Lengai Volcano, Tanzania: Melt inclusion study. *Lithos, 152*, 23–39. https://doi.org/10.1016/j.lithos.2012.01.021.

Shen, Q., Friese, K., Gao, Q., Yu, C., Kimirei, I. A., Kishe-Machumu, M. A., Zhang, L., Wu, G., Liu, Y., Zhang, J., Mgana, H., Dadi, T., Mpanda, D. W., & Chen, S. S. (2022). Status and changes of water quality in typical near-city zones of three East African Great Lakes in Tanzania. *Environmental Science and Pollution Research, 29*(23), 34105–34118. https://doi.org/10.1007/s11356-021-18079-w.

Shinya, W. M. (1998). Canada's new minerals and metals policy: advancing the concept of sustainable development in the minerals and metals industry. *Resources Policy, 24*(2), 95–104. https://doi.org/10.1016/S0301-4207(98)00013-0.

Simate, G. S. (2021). Legislation and policies governing the management of acid mine drainage. In G. S. Simate, & S. Ndlovu (Eds.), *Acid mine drainage: From waste to resources* (pp. 73–96). CRC Press. https://www.taylorfrancis.com/chapters/edit/10.1201/9780429401985-5/legislation-policies-governing-management-acid-mine-drainage-geoffrey-simate.

Spirnak, R., Shafer, M., Risch, C., O'Neal, M., & Ziemkiewicz, P. (2025). Evaluation of watershed-scale acid mine drainage treatment in the Muddy Creek Watershed, West Virginia. *Mine Water and the Environment, 44*(3), 579–594. https://doi.org/10.1007/s10230-025-01052-1.

Tang, Z., & Qin, D. (2024). Sustainable mining and the role of environmental regulations and incentive policies in BRICS. *Resources Policy, 90*, 104718. https://doi.org/10.1016/j.resourpol.2024.104718.

UNEP. (2020). Global industry standard on tailings management. *United Nations Environment Programme (UNEP)*. https://www.unep.org/resources/report/global-industry-standard-tailings-management.

Vice President's Office. (2021). *National Environmental Policy, 2021*. https://www.vpo.go.tz/publications/policies.

van Driel, M., Biermann, F., Kim, R. E., & Vijge, M. J. (2024). The impact of the Sustainable Development Goals on global policies on sustainable consumption and production. *Globalizations, 21*(8), 1366–1382. https://doi.org/10.1080/14747731.2024.2351301.

Verburg, R., Bezuidenhout, N., Chatwin, T., & Ferguson, K. (2009). The global acid rock drainage guide (GARD Guide). *Mine, Water and the Environment, 28*(4), 305–310. https://doi.org/10.1007/s10230-009-0078-4.

Wang, J., Wei, X., & Guo, Q. (2018). A three-dimensional evaluation model for regional carrying capacity of ecological environment to social economic development: Model development and a case study in China. *Ecological Indicators, 89*(January), 348–355. https://doi.org/10.1016/j.ecolind.2018.02.005.

Wang, H., Li, T., Zhu, J., Jian, Y., Wang, Z., & Wang, Z. (2023). China's new environmental protection law: Implications for mineral resource policy, environmental precaution and green finance. *Resources Policy, 85*, 104045. https://doi.org/10.1016/j.resourpol.2023.104045.

WHO. (2013). *Guidelines for the safe use of wastewater, excreta and greywater - Volume 4*. World Health Organisation (WHO). https://www.who.int/publications/i/item/9241546859.

Wiesmeth, H. (2021). Market-oriented policy tools. In *Implementing the circular economy for sustainable development* (pp. 183–193). Elsevier. https://doi.org/10.1016/B978-0-12-821798-6.00016-8.

Wood, M. H. (2009). The Seveso II experience in the application of generic substance criteria to identify major hazard sites. *Journal of Hazardous Materials, 171*(1–3), 16–28. https://doi.org/10.1016/j.jhazmat.2009.05.110.

Word Bank. (2024). *Achieving sustainable and inclusive artisanal and small-scale mining (ASM): A renewed framework for world bank engagement*. https://doi.org/10.1596/42156.

Zhang, Z., Hu, B., & Qiu, H. (2022). Comprehensive evaluation of resource and environmental carrying capacity based on SDGs perspective and three-dimensional balance model. *Ecological Indicators, 138*, 108788. https://doi.org/10.1016/j.ecolind.2022.108788.

Index

Page numbers followed by "*f*" and "*t*" indicate, figures and tables respectively.

L

M

N

O

P

R

S

[illegible]
[illegible]
[illegible]

Printed and bound by CPI Group (UK) Ltd, Croydon, CR0 4YY
14/09/2026
02216291-0001